FRANK SCHAFFER'S
CHEMISTRY
FOR EVERYDAY

Written by David Thurlo

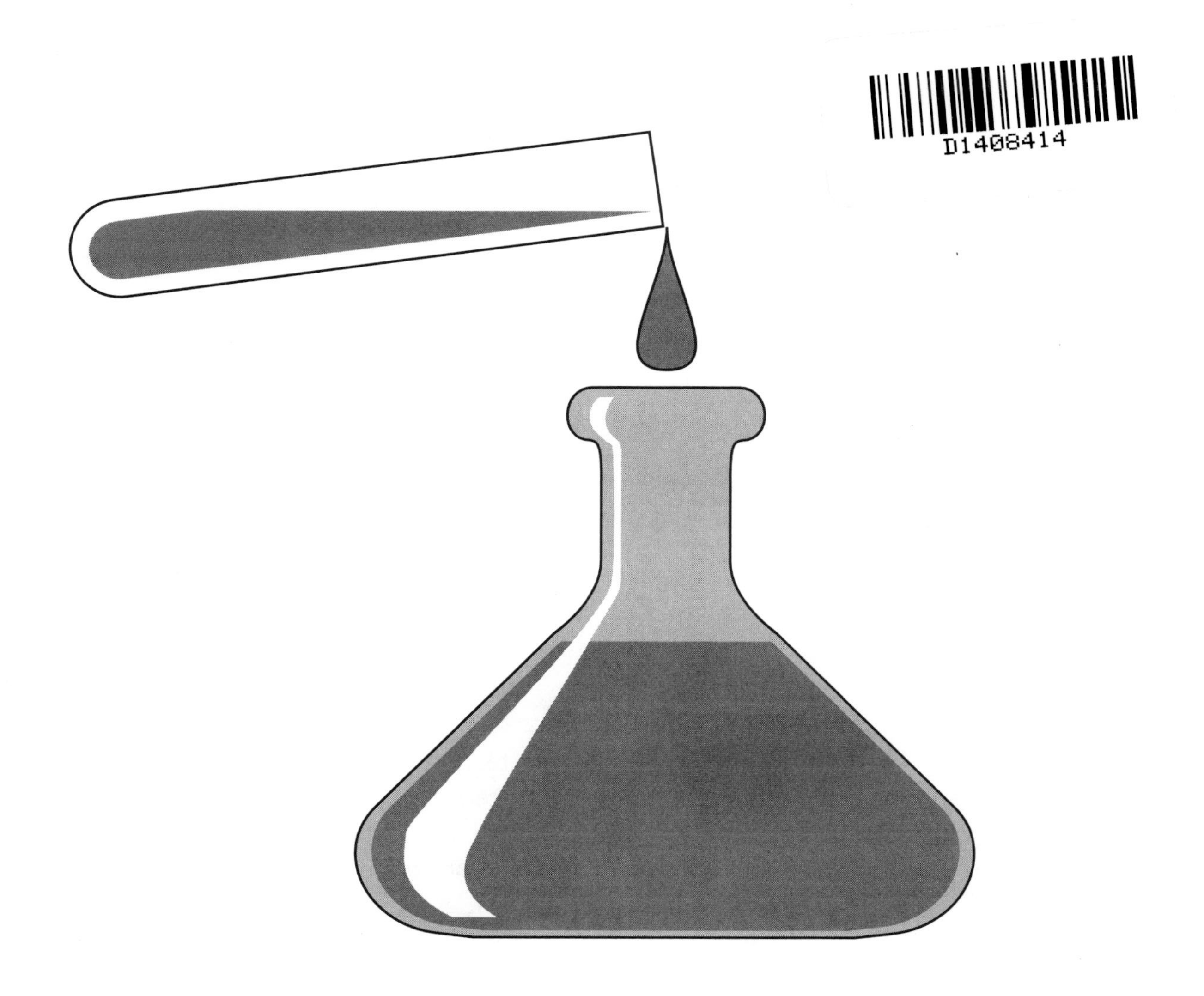

Copyeditors: Cindy Barden and Doug Fledderjohn

Interior Design: Good Neighbor Press, Inc.

Cover Illustration: Kent Publishing

FS110623 Frank Schaffer's Chemistry for Everyday

23740 Hawthorne Blvd.
Torrance, CA 90505

Table of Contents

Introduction

Frank Schaffer's Chemistry for Everyday contains a wealth of interesting activities students can complete to learn more about important topics in chemistry. Topics covered in this book include scientific methods applied to chemistry, determining the specific gravity, explaining states of matter, balancing chemical equations, the Periodic Table of the Elements (looking at the groups, periodic properties, electronic configurations), covalent and ionic bonds, and oxidation-reduction reactions.

Easy to use and self-explanatory, most of these reproducible activities can be completed by students at their desks using textbook references. A few require access to materials like encyclopedias and magazines usually found within the classroom. Often activities reinforce each other, allowing more in-depth work, some of which may be done at home. Several hands-on activities require students to carry out simple investigations using inexpensive equipment and supplies usually found in science storerooms.

Most of the activities (crossword puzzles, vocabulary matching, chart making, and problem solving) may be conducted individually, though a few requiring math skills may benefit from an introduction or preview by the teacher. Many of the activities are suitable for homework and/or non-instructional time, where students take responsibility for carrying out the tasks without assistance, keeping in mind their responsibility to carry out the work safely.

Whether you use these materials to supplement your chemistry class, give students an opportunity to reinforce their basic knowledge and scientific skills, or just make chemistry more interesting for the learner, you'll find the chemistry activities in this book useful and relevant to your objectives.

Chemistry References

Anthony C. Wilbraham, Dennis D. Staley, Candace J. Simpson, Michael S. Matta, *Chemistry, Third Edition* (Menlo Park, CA, Addison-Wesley, 1993)

Anthony C. Wilbraham, Dennis D. Staley, Candace J. Simpson, Michael S. Matta, *Chemistry, Second Edition* (Menlo Park, CA, Addison-Wesley, 1987)

Sargent-Welch, *Periodic Table of the Elements*, Catalog Number S-18806 (Skokie, IL, Sargent-Welch Scientific, 1968)

Denise Eby, Robert B. Horton, *Physical Science* (New York, NY, Macmillan Publishing, 1986)

Anthea Maton, et al, Matter, *Building Block of the Universe* (Englewood Cliffs, NJ, Prentice Hall, 1993)

Richard S. Burington, *Handbook of Mathematical Tables and Formulas* (Sandusky, Ohio, Handbook Publishers, Inc., 1946)

John R. Holum, *Elements of General and Biological Chemistry*, Second Edition (New York, NY, John Wiley and Sons, 1968)

David Thurlo, *Semiconductor Learning Materials* (Albuquerque, NM, Signetics Corporation, 1984)

Name ______________________ Date ______________

FRANK SCHAFFER'S **CHEMISTRY FOR EVERYDAY**

Reading the Periodic Table of the Elements

The modern arrangement of the elements in the Periodic Table of the Elements places them into rows and columns based upon groups with similar properties and characteristics. Learning to read this periodic chart makes it easier to learn and study the elements, and therefore, chemistry.

Materials Needed: A Periodic Table of the Elements

Using a periodic table, answer the following questions:

1. Find the names of the elements with the following chemical symbols:

   a) H ______ b) Al ______ c) Fe ______
   d) Eu ______ e) Lr ______ f) Cs______
   g) Kr ______ h) Pt ______ i) Xe ______

2. Find the chemical symbols for the following elements:

   a) helium ______ b) iodine ______ c) plutonium ______
   d) gold ______ e) lead ______ f) tin ______
   g) uranium______ h) einsteinium______
   i) carbon______ j) chlorine ______

3. Find the atomic numbers for the elements in Question 1.

   a) H ______ b) Al ______ c) Fe ______
   d) Eu ______ e) Lr ______ f) Cs______
   g) Kr ______ h) Pt ______ i) Xe ______

4. Find the atomic mass (weight) for the elements in Question 2.

   a) helium ______ b) iodine ______ c) plutonium ______
   d) gold ______ e) lead ______ f) tin ______
   g) uranium______ h) einsteinium______
   i) carbon______ j) chlorine ______

5. Elements in the Periodic Table of the Elements are arranged in columns or groups according to their properties. Name four elements that have properties similar to hydrogen.

   ______ ______ ______ ______

6. Name two elements that have properties similar to copper.

   ______ ______

7. Helium is a gas that will not burn. Name three other gaseous elements that probably will not burn either.

   ______ ______ ______

8. The elements in the Periodic Table are arranged from left to right and top to bottom based upon what value?

   ____

Name ____________________ Date ____________

FRANK SCHAFFER'S **CHEMISTRY FOR EVERYDAY**

Periodic Table Crossword

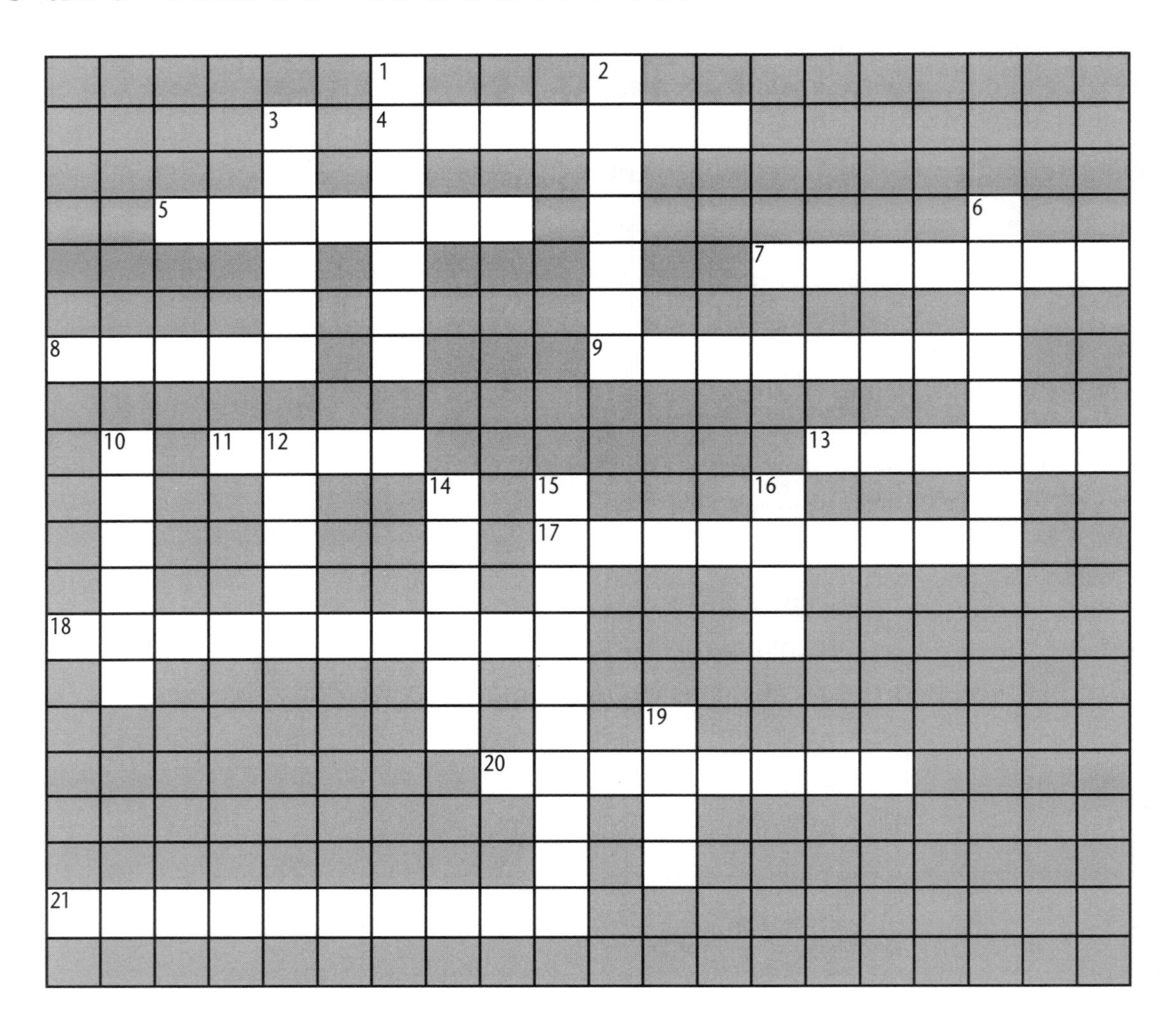

Across

4. the atomic weight is 192.2
5. the atomic weight is 158.9
7. the atomic weight is 183.8
8. the atomic weight is 39.9
9. the lightest element
11. the atomic weight is 55.8
13. the atomic number is 56
17. the atomic weight is 101
18. the atomic number is 43
20. the atomic number is 24
21. has an atomic number of 103

Down

1. has the atomic symbol Si
2. the atomic number is 83
3. has the atomic number 6
6. the atomic number is 87
10. the atomic symbol is Cu
12. has the atomic number 86
14. has the atomic symbol Cs
15. the atomic weight is 145
16. the atomic weight is 20.1
19. has an atomic weight of 196.9

Scientific Method Applied to Chemistry

There are many variations of the scientific process called the scientific method. Each is a sequence of events and activities that serves as a guide for problem solving and investigations. Many discoveries in the field of chemistry came about through use of the scientific method. Demonstrate your understanding of the scientific method by describing the various steps taken that are usually included in this process.

1. Objective:

2. Research:

3. Hypothesis:

4. Experiment:

5. Observations:

6. Data Collection:

7. Results:

8. Conclusion:

9. Verification:

Name ______________________________ Date ____________

Determining Physical Properties

Materials: chemistry textbook and/or Periodic Table of the Elements chart

We are surrounded by all forms of matter. Matter is anything that has mass and takes up space. This includes the ground below us, the air we breathe, and the paper these words are written on. All matter can be grouped according to its physical properties—those qualities or conditions that can be observed and measured without changing the composition of the matter.

Use your chemistry textbook and/or a comprehensive Periodic Table of the Elements to determine the physical properties of the substances listed below. Complete Table A.

Table A

Substance	State at Room Temp. (solid, liquid, gas)	Melting Point	Boiling Point	Other
1. water				
2. iron				
3. nitrogen				
4. oxygen				
5. copper				
6. sulfur				
7. graphite				
8. chlorine				
9. gold				
10. platinum				
11. silver				

12. What are some other physical properties of matter not included on the chart above?

Name ______________________________ Date ______________

State Properties

The three physical states of matter are properties that help distinguish one substance from another. Compare the three states of matter by completing Chart A below.

Use terms such as fixed, variable, non-variable, small, medium, large, barely, yes, definite, or indefinite as entries in the chart.

Chart A

Property	Solid	Liquid	Gas
1. Mass			
2. Shape			
3. Volume			
4. Temperature increase and volume			
5. Ability to be compressed			
6. Motion of molecules			
7. Examples at room temperature			

8. Make a sketch representing molecules of water at all three states of matter.

Name ______________________________ Date ______________

Understanding Matter and Energy

The following terms are important when studying matter, energy, and chemical change. Explain, describe, or define them. Give examples for each.

1. chemical change

2. chemical property

3. chemical reaction

4. chemical symbol

5. kinetic energy

6. law of conservation of energy

7. law of conservation of mass

8. physical change

9. potential energy

10. scientific law

Name ______________________ Date ______________

Basic Terms in Chemistry

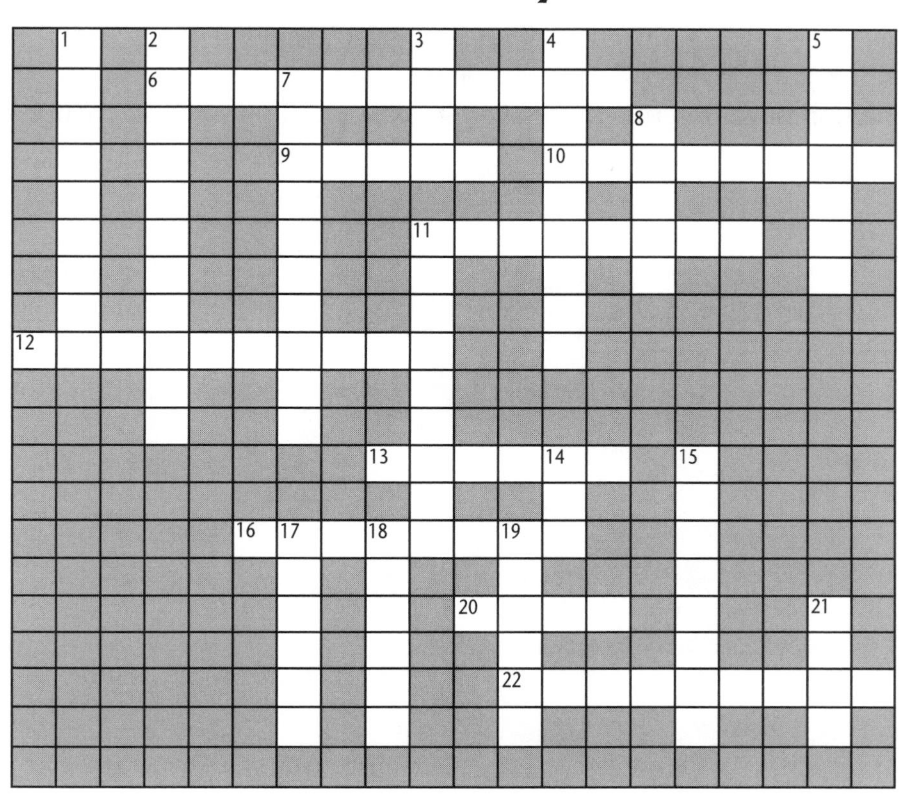

Across

6. noting and recording facts and descriptions
9. any part of a system with uniform composition and properties
10. the new substances formed from a chemical reaction
11. a homogeneous mixture
12. a descriptive model based upon a prediction
13. the capacity for doing work
16. simplest forms of matter that can exist under normal lab conditions
20. energy transferred because of a temperature difference
22. the starting substances of a chemical reaction

Down

1. the study of the composition of substances and their interactions
2. uniform in composition
3. the amount of matter something contains
4. substances composed of more than one element
5. a physical blend of two or more substances
7. a test to see if a hypothesis is correct
8. matter that has definite shape and volume
11. a particular kind of matter that has a uniform and definite composition
14. matter that takes the shape and volume of its container
15. a quality or condition of a substance
17. matter that has a fixed volume and takes the shape of its container
18. anything that has mass and takes up space
19. a tested model that explains certain results from experiments
21. observations recorded from an experiment

Name ______________________ Date ____________

Recognizing Physical and Chemical Changes

One of the most basic concepts of chemistry is understanding the difference between chemical and physical changes.

Write a complete definition of each term. Then give ten examples of a physical change and ten examples of a chemical change.

1. Physical change

   Examples:

   a.

   b.

   c.

   d.

   e.

   f.

   g.

   h.

   i.

   j.

2. Chemical change

   Examples:

   a.

   b.

   c.

   d.

   e.

   f.

   g.

   h.

   i.

   j.

Name ______________________ Date ____________

Systems of Measurement

The system commonly in use today among scientists and students of science is the International System of Units, abbreviated as SI. There are seven basic units.

Complete the chart below for the SI system, listing the quantity measured, the unit, and the SI symbol for each of the seven basic units. The first one is done for you.

Quantity	Unit	SI symbol
1. Length	meter	m
2. Mass		
3. Time		
4. Electric current		
5. Thermodynamic temperature		
6. Amount of substance		
7. Luminous intensity		

Other units used in scientific investigations are more commonly used in the lab. Fill in the chart below.

Quantity	Name and relationship	symbol
8. Volume		
9. Mass		
10. Density		
11. Temperature		
12. Time		
13. Pressure		
14. Energy		

Name ______________________________ Date ______________

Measuring Up

One of the most important skills to master in chemistry, as in so many other activities in school or at work, is measuring accurately. Demonstrate how you measure up to the challenge by answering the following questions:

1. What is the difference between quantitative and qualitative measurements?

2. What are some of the important contributions Antoine-Laurent Lavoisier made to the science of chemistry?

3. Explain the difference between accuracy and precision. Give an example.

4. What is a significant figure? How is it determined?

5. How are significant figures used in calculations?

Name ______________________ Date ____________

Measuring Metrically with Meters

The dimensions of length, width, and depth are all determined in the SI system with the basic unit, the meter. Complete Table A, giving the units commonly used in the metric system to express length. Start with the largest unit at the top of the chart.

Table A

Unit	Symbol	Relationship
1.		
2.		
3.		
4.		
5.		
6.		
7.		

On Table B, enter the prefixes, symbols, and meanings for the common prefixes used in the SI system for length.

Table B

Prefix	Symbol	Meaning
8.		
9.		
10.		
11.		
12.		
13.		

Name ______________________________ Date ______________

Specific Gravity Specifics

One of the physical properties of a substance is its specific gravity, which is the density of the substance compared to the density of another reference substance at the same temperature and pressure. Water is usually the reference substance. The formula for determining the specific gravity is:

$$\textbf{Specific gravity} = \frac{\textbf{density of substance (g/cm}^3\textbf{)}}{\textbf{density of water (g/cm}^3\textbf{)}}$$

Materials needed: A detailed Periodic Table of the Elements or a reference chart of the elements, and a calculator

Complete Table A, finding the symbols, densities at STP (standard temperature and pressure of 0°C and 1 atmosphere), and calculating the specific gravity of the listed elements. Specific gravity is expressed as a number, with no units. The density of water at 0°C is 0.917 g/cm^3.

Table A

Element	Symbol	Density at STP	Specific Gravity
1. aluminum			
2. beryllium			
3. carbon			
4. iron			
5. sodium			
6. titanium			
7. chromium			
8. copper			
9. silver			
10. gold			
11. barium			
12. uranium			

Name ______________________ Date ____________

FRANK SCHAFFER'S **CHEMISTRY FOR EVERYDAY**

Problems With Numbers?

Significant figures in a measurement include all the digits that can be known accurately, plus a last digit that must be estimated. Rules for determining significant figures include the following:

- All non-zero digits are significant.
- Zeros appearing between non-zero digits are significant.
- Zeros in front of all non-zero digits are not significant.
- Zeros at the end of a number and to the right of the decimal point are significant.

Find the number of significant figures for each number below.

1. 142 ____________
2. 0.073 ____________
3. 1.071 ____________
4. 10,810 ____________
5. 5.00 ____________
6. 55.320 ____________
7. 1.010 ____________
8. 154 ____________
9. 8,710 ____________
10. 1.0004 ____________

Round off the following numbers to three significant figures.

11. 88.473 ____________
12. .00086321 ____________
13. 67.048
14. 8,500 ____________
15. 12.17 ____________

Express the following numbers in scientific notation (exponential) form.

16. 1200 ____________
17. 25,000 ____________
18. 81,000,000 ____________
19. .007 ____________
20. .0035 ____________

Express these exponential numbers as whole numbers.

21. 3×10^5 ____________
22. 2.3×10^2 ____________
23. 6×10^{-3} ____________
24. 2.7×10^{-4} ____________
25. 6.1×10^6 ____________

Name ______________________ Date ____________

FRANK SCHAFFER'S **CHEMISTRY FOR EVERYDAY**

Determining Specific Gravity

One of the physical properties of a substance, whether an element or compound, is specific gravity. Geologists use this property as an aide in identifying minerals, which are chemical compounds or individual elements found naturally. In this activity, you will determine the specific gravity of ten different minerals by comparing their weight in air with the weight of an equal volume of water. The process is simple. The mineral is weighed with a spring scale when dry, then submerged completely in a container of water and weighed again. The difference in dry weight and weight in water is equal to the amount of water displaced, which is a volume of water equal to the volume of the mineral.

Using the formula, specific $\textbf{gravity} = \dfrac{\textbf{weight of sample in air}}{\textbf{weight of equal volume of water}}$

one can determine the specific gravity of any non-permeable mineral.

Materials: spring scale, fine string or thread, minerals, beaker of water, calculator

Procedure:

1. Weigh each mineral. (Tie string to the mineral and suspend it from the spring scale). Record your data on Table A, under Dry Weight.
2. Weigh each mineral in the water by suspending it as before, then lowering the mineral into the water so it is completely submerged, but not touching the sides or bottom of the beaker. Record that data under Weight in Water.
3. Find the difference between the dry weight and weight in water for each mineral. Record that data under Weight of Equal Volume of Water.
4. Calculate the specific gravity by dividing the dry weight by the difference entered under Weight of Equal Volume of Water. Record the specific gravity for each mineral.

Mineral	Dry Weight	Weight in Water	Weight of Equal Volume of Water	Specific Gravity

Name ______________________________ Date ______________

Atomic Structure

Understanding the fundamental structure of atoms is a necessary skill for understanding elements, compounds, and chemical reactions. Chemical changes take place at the level of individual atoms. Knowing about electrons and atomic structure allows one to predict the outcome of experiments.

Demonstrate your knowledge of basic atomic structure by matching the term with the appropriate definition or description.

Definitions or Descriptions

1. ____ The smallest particle of an element that retains the properties of that element
2. ____ Established that atoms of one element are not changed into atoms of another element in a chemical reaction
3. ____ Negatively charged subatomic particle
4. ____ A low pressure gas-filled glass tube which generates a glowing beam between cathode and anode
5. ____ A positively charged subatomic particle found in the nucleus
6. ____ A subatomic particle with no charge
7. ____ Composed of neutrons and protons
8. ____ The number of protons in the nucleus of the atom of that element
9. ____ One twelfth the mass of a carbon atom that contains 6 protons and 6 neutrons
10. ____ The total number of protons and neutrons in the nucleus
11. ____ Atoms that have the same number of protons but different numbers of neutrons
12. ____ The weighted average of the masses of the isotopes of that element

A. electron	E. nucleus	I. atomic mass unit
B. isotope	F. proton	J. atomic number
C. mass number	G. atom	K. cathode ray
D. neutron	H. atomic mass	L. Dalton's atomic theory

Name ______________________ Date ______________

FRANK SCHAFFER'S CHEMISTRY FOR EVERYDAY

Using the Periodic Table of the Elements

The elements are arranged in rows and columns on the periodic table according to similarities in their properties. Their chemical properties are a result of their electron configuration and it is important to know how to gain information from the periodic table about each element. One of the first steps in reading the periodic table is discovering how to find an element, and another is learning to interpret the numbers and symbols there.

Use the Periodic Table of the Elements to complete the chart below.

Element	# of protons	mass # (rounded)	electron #	Atomic #	# of neutrons
1. hydrogen					
2. helium					
3. carbon					
4. nitrogen					
5. oxygen					
6. cobalt					
7. copper					
8. iron					
9. uranium					
10. krypton					

Name ______________________________ Date ______________

FRANK SCHAFFER'S **CHEMISTRY FOR EVERYDAY**

Problems With the Atomic Structure

Demonstrate your understanding of the principles of atomic structure by correctly answering the questions below.

1. When looking at the period table of the elements, why are the atomic masses of the elements usually not whole numbers?

2. How does the relative size and density of an atom compare to its nucleus?

3. Describe the isotopes of hydrogen.

4. John Dalton was one of the pioneers of atomic theory. He made important contributions to explaining the structure of elements and how they combine to form compounds. There are some aspects of his theories that we know today were incorrect. What mistakes did Dalton make?

5. Draw and label a simple cathode ray tube, explain how it works, then identify what forces are capable of bending or deflecting a cathode ray.

Going Deeper: The Chemistry of Metal Production

Write a report describing, in detail and in your own words, the modern manufacturing process for the production of one of the following metals—aluminum, copper, steel, or zinc.

Name ______________________________ Date ______________

Explaining Chemical Terms

The language of chemistry includes many important terms. If students understand their meanings and relationships, it is much easier to understand the concepts.

Explain or define the following terms related to chemical names and formulas.

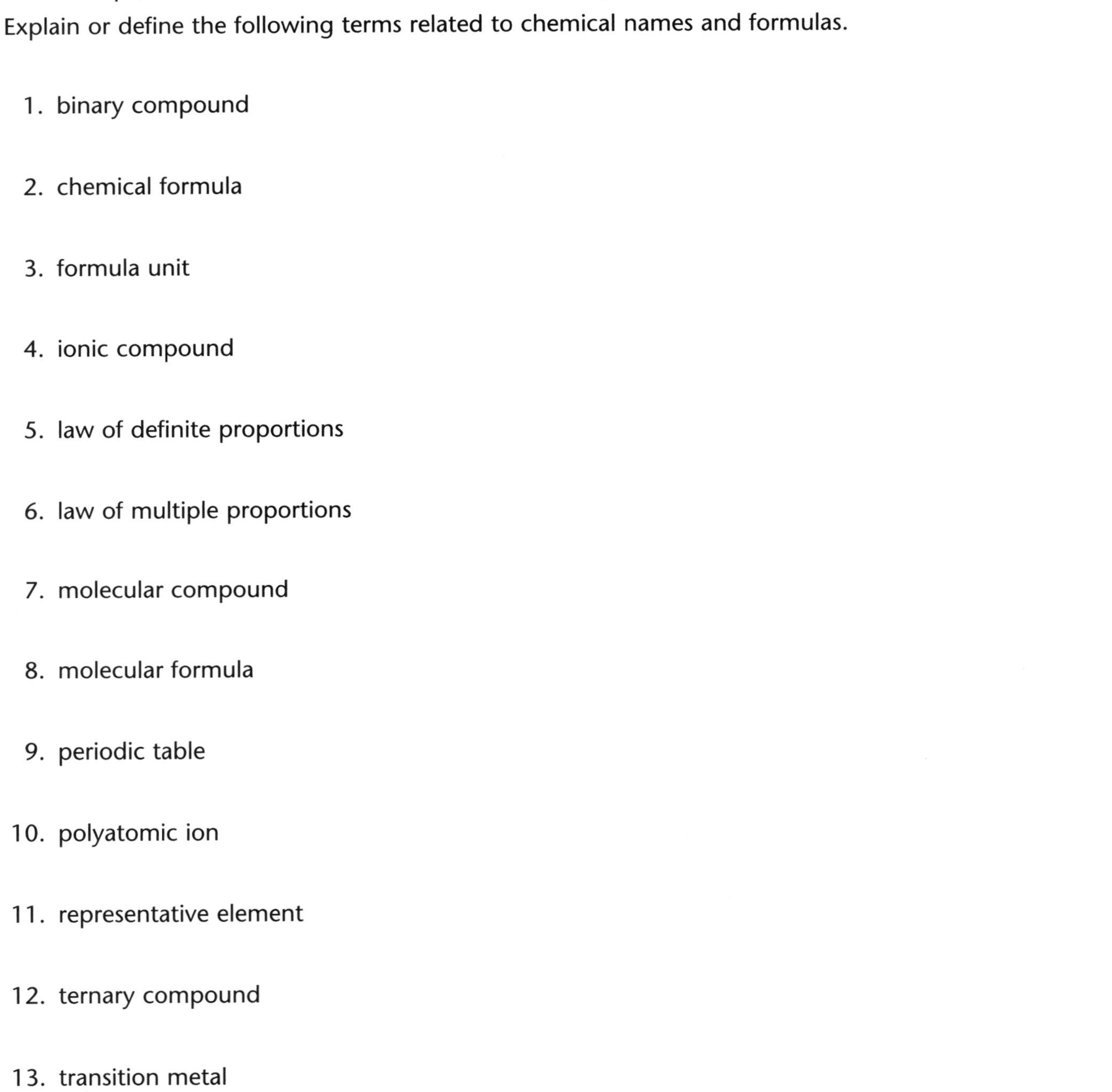

1. binary compound
2. chemical formula
3. formula unit
4. ionic compound
5. law of definite proportions
6. law of multiple proportions
7. molecular compound
8. molecular formula
9. periodic table
10. polyatomic ion
11. representative element
12. ternary compound
13. transition metal

Name ______________________ Date ______________

Formulas and Formula Names

Learning how to write the formulas for chemical compounds, and how to read these formulas, is based upon several important rules. The process can seem complicated and confusing at times, but can be learned with practice. Try out your skills by answering the following:

Write the chemical formulas for the following:

1. barium sulfate ______________
2. aluminum hydrogen carbonate ______________
3. sodium hypochlorite ______________
4. lead (IV) chromate ______________
5. aluminum hydroxide ______________
6. potassium chlorate ______________
7. zinc hydrogen sulfate ______________
8. carbon disulfide ______________
9. perchloric acid ______________

Write names for these chemical formulas.

10. AlI_3 ______________
11. FeO ______________
12. Cu_2S ______________
13. $CaSe$ ______________
14. $Mg_3(PO_4)_2$ ______________
15. Li_2CrO_4 ______________
16. K_2SiO_3 ______________
17. LiF ______________

Name ______________________ Date ______________

Models of Molecular Compounds

A molecular formula shows the number and kinds of atoms in a molecule of a given compound. A simple model of the molecular structure of these compounds can be shown with different size or colored circles to represent each atom.

Using larger circles to represent carbon, medium-sized circles to represent oxygen, and small circles to represent hydrogen, draw models of the following molecular compounds that are given as molecular formulas. Then name the compounds. Identify each atom with a C, H, or O, or use colored pencils and a color key.

Molecular formula	Drawing of molecule	Name of compound
1. H_2O		
2. CO		
3. CO_2		
4. C_2H_6		
5. C_2H_6O		
6. CH_2O_2		

Name ______________________________ Date ______________

Names and Formulas

The process of learning to identify each chemical substance and formula by name and composition is a long and complicated one, but it is important to know how to "speak" chemistry in order to understand the processes.

Answer the following questions about formulas and names.

1. Explain how certain ionic charges can be determined by using the periodic table.

2. Explain the old system of naming ions from the transition metals.

3. What is the Stock name?

4. How can you recognize most polyatomic ions by their names?

5. What is the rule for writing the Stock names of ionic compounds?

6. Why are Roman numerals sometimes used in naming binary ionic compounds?

Name ______________________ Date ____________

FRANK SCHAFFER'S CHEMISTRY FOR EVERYDAY

Terms Related to Chemical Quantities

Chemists and chemistry students analyze the composition of substances and do chemical calculations relating the quantities of reactants and products in chemical reactions. In order to carry out any work concerning the chemical nature of a substance or a reaction, you need to know how to measure chemicals accurately and consistently.

Define or describe the following chemical terms.

1. Avogadro's number
2. empirical formula
3. gram atomic mass
4. gram formula mass
5. gram molecular mass
6. molar mass
7. molar volume
8. mole
9. percent composition
10. representative particle
11. standard temperature and pressure

Name ______________________ Date ______________

FRANK SCHAFFER'S **CHEMISTRY FOR EVERYDAY**

Key Terms Used in Studying Chemical Reactions

Chemical reactions are usually explained by scientists and students in the form of chemical equations, a convenient shorthand listing the reactants and products.

Explain or define the terms listed below. Give an example of each.

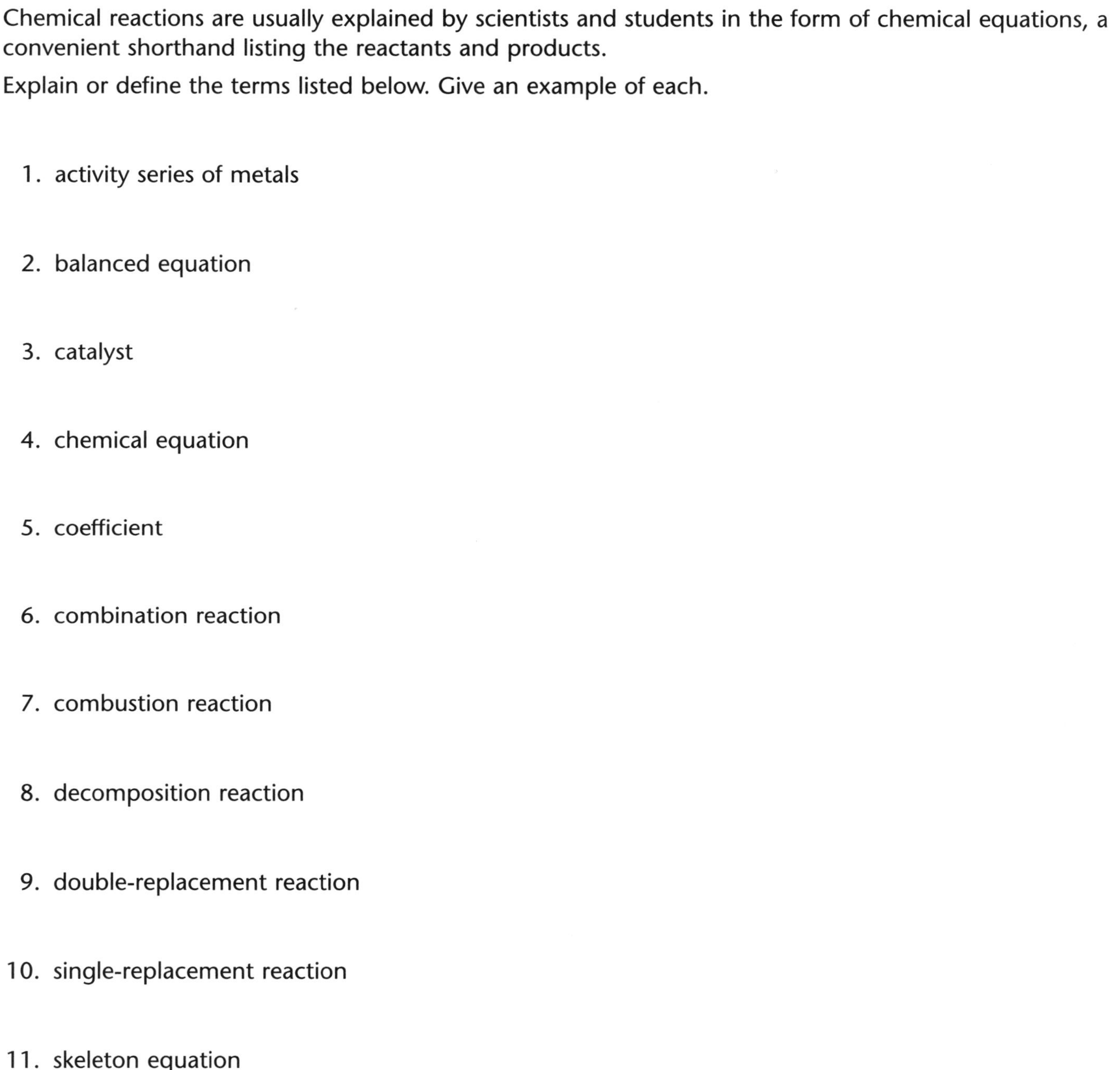

1. activity series of metals
2. balanced equation
3. catalyst
4. chemical equation
5. coefficient
6. combination reaction
7. combustion reaction
8. decomposition reaction
9. double-replacement reaction
10. single-replacement reaction
11. skeleton equation

Name ______________________________ Date ______________

Understanding Chemical Symbols

In addition to the symbols for the chemicals themselves, certain symbols are used when writing chemical equations. These symbols designate a form of a chemical, the inclusion of a catalyst, the release of energy, and so forth.

Listed below are the explanations, definitions, or descriptions for common chemical symbols. Use your chemistry text to discover which symbol or symbols match these explanations. Then draw the symbol to the right of that explanation.

Symbol	Explanation
1. ______________	alternative to (s), and indicates a solid product or precipitate
2. ______________	alternative to (g), used to indicate a gaseous product
3. ______________	this position indicates its function as a catalyst
4. ______________	sometimes used as an alternative to a right facing arrow
5. ______________	a reactant or product in the solid state
6. ______________	a reactant or product in the liquid state
7. ______________	an aqueous solution, the substance is dissolved in water
8. ______________	a reactant or product in the gaseous state
9. ______________	used to separate two reactants or two products
10. ______________	means "yields" and separates reactants from products
11. ______________	used when a reaction is reversible
12. ______________	used to indicate heat is added to the reaction

Name ______________________________ Date ______________

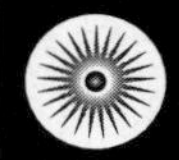

"Fleshing Out" Skeleton Equations

Skeleton equations are chemical equations that show the reactants and products in a chemical reaction, but do not indicate their relative amounts.

Chemical reactions can also be described with words instead of equations, but still, the relative amounts are not always clear. When a chemical equation is adjusted with the careful addition of coefficients, so that each side of the equation has the same number of atoms of each element, the equation is balanced.

Demonstrate your understanding of chemical reactions and equations by completing the following:

1. Give ten examples of skeleton equations that are not balanced.

   Example: $Fe + O_2 \rightarrow Fe_2O_3$

   a.) f.)

   b.) g.)

   c.) h.)

   d.) i.)

   e.) j.)

2. Write sentences that describe the ten skeleton equations in Number 1.

   Example: Iron reacts with oxygen to form iron oxide.

   a.) f.)

   b.) g.)

   c.) h.)

   d.) i.)

   e.) j.)

3. Rewrite the chemical equations in Number 1 as balanced equations.

   Example: $4\ Fe + 3O_2 \rightarrow 2\ Fe_2O_3$

   a.) f.)

   b.) g.)

   c.) h.)

   d.) i.)

   e.) j.)

Name ______________________________ Date ______________

FRANK SCHAFFER'S **CHEMISTRY FOR EVERYDAY**

Balancing Chemical Equations

In order to balance a chemical equation, one must follow certain rules:

1. Make sure the formulas are correct for all the products and reactants.
2. Make sure the reactants are on the left side of the arrow and that a plus sign separates individual reactants and products.
3. Count the number of atoms of each element on both sides of the equation. A polyatomic ion that is unchanged on both sides of the equation counts as a single unit, as if it were an element.
4. Balance the elements one at a time using coefficients. Subscript numbers cannot be changed.
5. Make sure all the coefficients are in the lowest possible ratio.

Practice by writing the balanced equation for each skeleton equations.

Skeleton equations	**Balanced equations**
1. $Na + Cl_2 \rightarrow NaCl$	______________
2. $Ag + S \rightarrow Ag_2S$	______________
3. $Mg + O_2 \rightarrow MgO$	______________
4. $Al_2O_3 \rightarrow Al + O_2$	______________
5. $H_2 + N_2 \rightarrow NH_3$	______________
6. $C + Fe_2O_3 \rightarrow Fe + CO$	______________
7. $KCl + F_2 \rightarrow KF + Cl_2$	______________
8. $NaBr + Ag_2SO_4 \rightarrow Na_2SO_4 + AgBr$	______________
9. $Fe_2O_3 + HCl \rightarrow FeCl_3 + H_2O$	______________
10. $Al + CuSO_4 \rightarrow Al_2(SO_4)_3 + Cu$	______________
11. $P + O_2 \rightarrow P_2O_5$	______________
12. $Al + H_2SO_4 \rightarrow Al_2(SO_4)_3 + H_2$	______________
13. $SrBr_2 + (NH_4)_2CO_3 \rightarrow SrCO_3 + NH_4Br$	______________
14. $C_2H_6 + O_2 \rightarrow CO_2 + H_2O$	______________
15. $Fe + O_2 \rightarrow Fe_3O_4$	______________

Name ______________________________ Date ______________

Stoichiometry: Calculating Quantities in Chemistry

Stoichiometry is the calculation of quantities in chemical equations. Calculations using balanced equations are called stoichiometric calculations.

Define or describe the following terms, which are all related to stoichiometry:

1. actual yield
2. endothermic reaction
3. enthalpy
4. excess reagent
5. exothermic reaction
6. heat of combustion
7. heat of reaction
8. limiting agent
9. percent yield
10. standard heat of formation
11. theoretical yield
12. thermochemical equation

Name ______________________________ Date ____________

Research Topics in Chemistry

Write a research paper on one of the following topics related to chemistry:

1. The chemistry of food products, from soft drinks to bread and cakes.
2. Providing evidence for criminal investigations, from fingerprinting to detecting trace amounts of poison.
3. The manufacture and use of Portland cement and other building products, such as concrete, glass, and special adhesives.
4. Chemistry in the medical professions, from pharmacy to the operating room.
5. Water as a chemical—how it is treated and used.
6. Common exothermic reactions in modern society.
7. The presence and consequences of man-made chemicals in our environment.
8. Uses of chemical products in industry and chemical production in your community.
9. Fertilizers and other chemicals used in agriculture.
10. Heroes of chemistry—their lives and discoveries.

Kinetic Theory and the Nature of Gases

Open a bottle of ammonia and the aroma soon reaches the noses of everyone in the room. This is because the molecules of ammonia have diffused from the bottle into the air and is evidence that molecules have motion.

Kinetic means *motion.* The kinetic theory states that the tiny particles in all forms of matter are in constant motion. For gases, this is especially true.

Discuss in detail the kinetic theory of gases, explaining their structure, motion, physical behavior, and relationship to temperature and pressure. Write your answers on another sheet of paper.

Name ______________________________ Date ______________

States of Matter: Coming to Terms

Match the following definitions or descriptions with the terms listed at the bottom of the page. Write the letter representing the term in the blank.

1. _______ The term for the conversion of a liquid to a gas or vapor below its boiling point.
2. _______ When vaporization of an uncontained liquid occurs.
3. _______ Produced when vaporized particles collide with the walls of a sealed container.
4. _______ The temperature at which the vapor pressure of the liquid is equal to the external pressure.
5. _______ The boiling point of a liquid at a pressure of 1 atm.
6. _______ The temperature at which a solid turns into a liquid.
7. _______ Forms when the particles are arranged into an orderly, repeating, three-dimensional pattern.
8. _______ The smallest group of particles within a crystal that retain the geometric shape of the crystal.
9. _______ Solids that lack an ordered internal structure.
10. _______ A form of amorphous solids, like glass.
11. _______ Occurs whenever the physical state of a substance changes.
12. _______ The change of a solid to a gas or vapor without passing through the liquid state.
13. _______ Heat required to melt one gram of a solid at its melting point.
14. _______ Amount of heat given up as one gram of liquid changes to a solid at the melting point.
15. _______ Number of calories required to change 1 g of a liquid to gas at the boiling point at atmospheric pressure.
16. _______ The heat released when 1 g of a gas condenses to a liquid at the boiling point.

A. amorphous solid
B. boiling point
C. crystal
D. evaporation
E. heat of condensation
F. heat of fusion
G. heat of solidification
H. heat of vaporization
I. melting point
J. normal boiling point
K. phase change
L. sublimation
M. supercooled liquid
N. unit cell
O. vapor pressure
P. vaporization

Name ______________________________ Date ______________

Explaining Liquids, Solids, and Plasmas

A gas is a very active state of matter, but the other states exhibit motion and distinct behavior and properties as well.

In the space below, explain the properties and characteristics of liquids, solids, and plasmas.

Liquids

Solids

Plasmas

Under Pressure

When moving bodies collide with each other, they generate a force on the object depending on their mass and velocity. This force is pressure.

Explain pressure and describe how it is measured. Make a simple sketch of a mercury barometer and explain how it works.

Name ______________________________ Date ______________

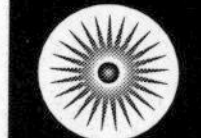

Solids and the Seven Crystal Systems

Most solid substances are crystalline in nature, the particles forming an orderly, repeating, three dimensional pattern. This pattern is called a crystal lattice. Each pattern has a regular shape, and the shape reflects the arrangement of the particles within the solid.

Crystals are classified into seven crystal systems with characteristic shapes. The shapes differ according to the angles between the faces and the number of edges of the faces that are equal. Make a sketch of each of the seven crystal systems and label which system each sketch represents.

Name ______________________________ Date ______________

Crystal Models

Color, cut out, and assemble these four crystal models using glue or tape. Identify what crystal system each model belongs to.

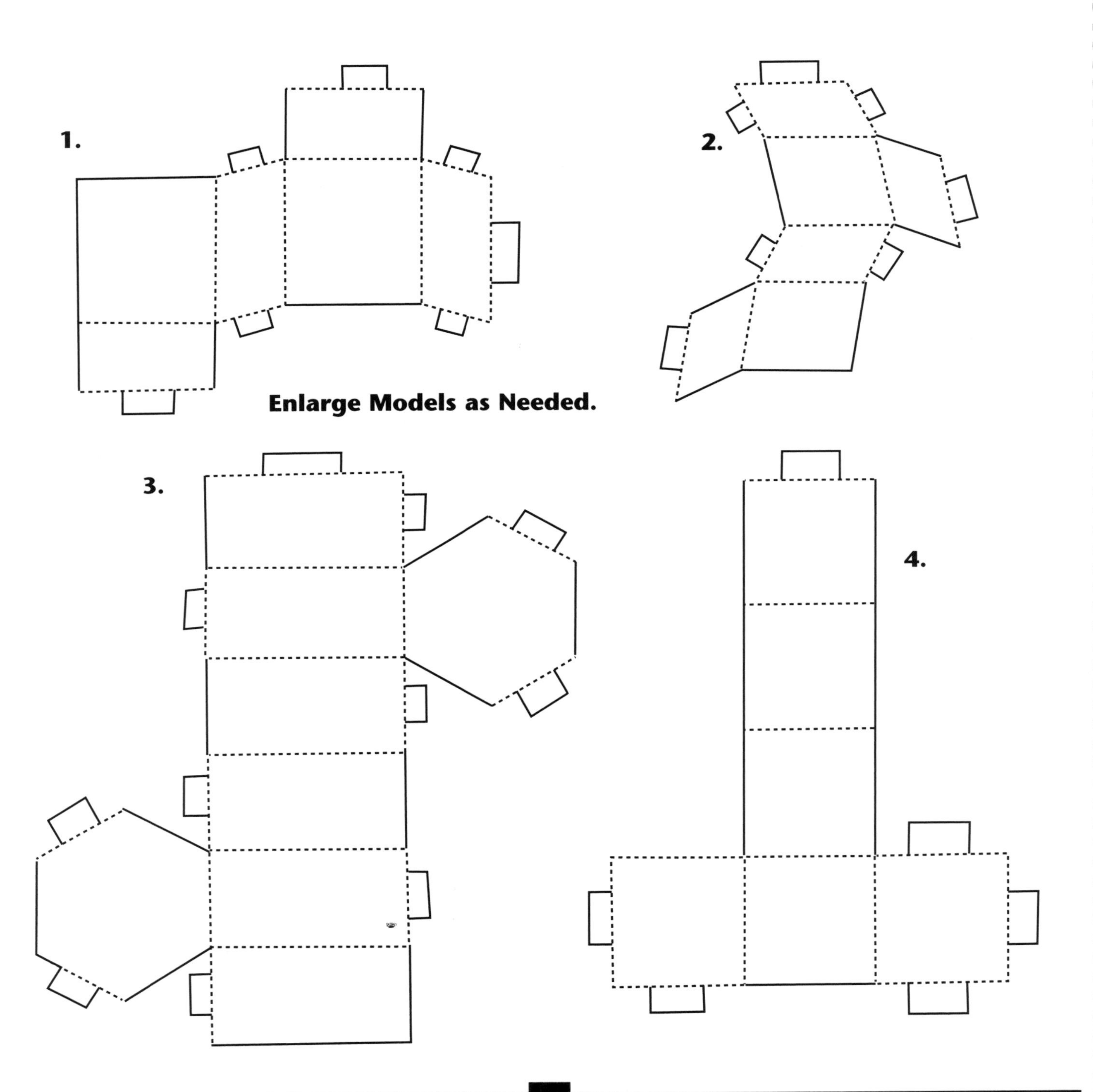

Name ______________________ Date ________

Air Force Demonstration

This activity, which demonstrates an important characteristic of gases could be conducted at home or in the lab.

Safety first: Make sure you follow directions carefully. Protect your skin and eyes from steam, hot water, and the heat source you are using.

Materials: a small kettle or large sauce pan, water, an oven mitt or tongs, and a 710 ml plastic water bottle with cap (drinking water is sold in these containers—which are available in grocery stores)

Procedure:

1. Place enough water in the kettle so the plastic water bottle, if pressed straight down into the water, will be ¾ immersed.
2. Heat the water in the kettle on a stove or hot plate to a temperature of 100 degrees celsius (boiling).
3. Turn off the heat. Press the empty, open water bottle (top side up) straight down into the hot water so that approximately ¾ of the bottle is under the water. Don't burn yourself and don't let any water get into the bottle.
4. Keep the bottle ¾ immersed for 70 seconds, then cap the bottle tightly and remove it from the hot water. Set the capped plastic bottle down on the table and observe. Answer the following questions.

1. What happens to the bottle?

2. What do you think caused the change?

3. Why was the bottle uncapped during the first part of the experiment?

4. What might have happened if the bottle had remained capped throughout the experiment?

5. Open the cap. What happens? Why?

6. Explain the results of this experiment in terms of temperature, volume, and pressure.

Name ______________________________ Date ______________

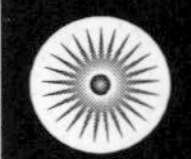

FRANK SCHAFFER'S **CHEMISTRY FOR EVERYDAY**

How Do Hot Air Balloons Work?

A hot air balloon is capable of rising and staying aloft because of certain properties of gases. Explain how and why a hot air balloon rises and stays aloft. Use a labeled sketch in your explanation.

The Quantum Mechanical Model of the Atom

In 1913, Niels Bohr, a Danish physicist, proposed a new model of the atom, in which he suggested that electrons are found in energy levels and that electrons can jump from one energy level to another by adding or losing a quantum of energy. Later, scientists improved on Bohr's model of the atom to eventually develop the modern theory, the quantum mechanical model.

Discuss the quantum model. Explain electron configurations, atomic emission spectrums, de Broglie's equation, and the Heisenberg uncertainty principle. Use a separate sheet of paper for your answers.

Name ______________________________ Date ______________

Even Gases Must Obey the Law

Several important observations about gases and their properties have led to rules they follow, and the results have been scientific laws. Explain or define the following laws, rules, and observed behavior of gases:

1. Boyle's law

2. Charles' law

3. combined gas law

4. Dalton's law of partial pressures

5. diffusion

6. effusion

7. Gay-Lussac's law

8. Graham's law of effusion

9. ideal gas constant

10. ideal gas law

11. partial pressure

Name ______________________________ Date ______________

Atoms and Electrons

The particle most responsible for the chemical properties of atoms is the electron. Understanding the structure and nature of matter at the atomic level is essential in explaining why elements and compounds behave as they do in chemical reactions.

The following terms are important when studying the properties of elements and the quantum mechanical model of the atom, which provides the modern description of electrons.

Define or describe the terms below:

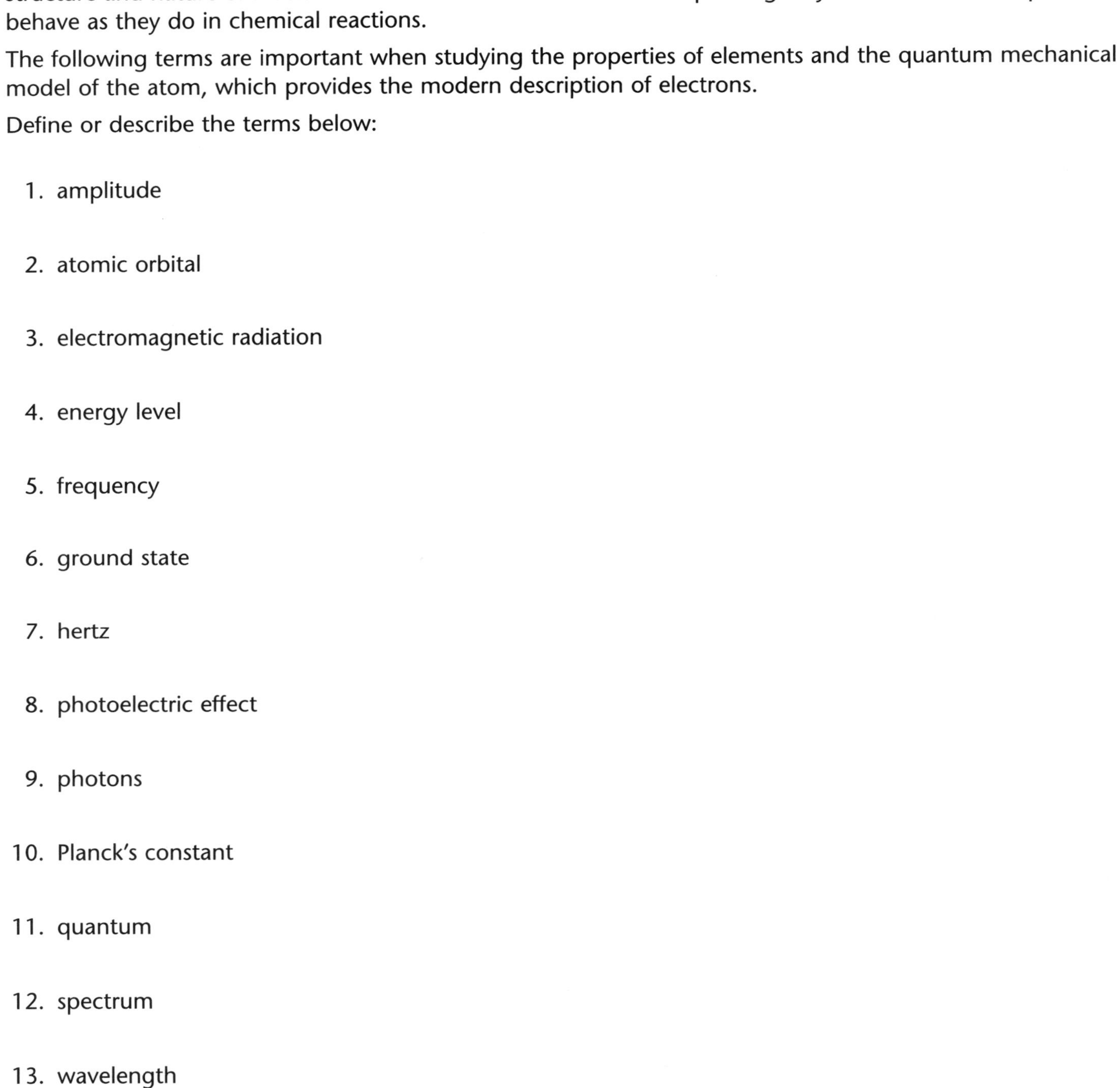

1. amplitude
2. atomic orbital
3. electromagnetic radiation
4. energy level
5. frequency
6. ground state
7. hertz
8. photoelectric effect
9. photons
10. Planck's constant
11. quantum
12. spectrum
13. wavelength

Name ______________________________ Date ______________

FRANK SCHAFFER'S **CHEMISTRY FOR EVERYDAY**

Don't Just Wave, Draw Waves

Sometimes a simple sketch is worth a thousand words of explanation.

1. Prove this by creating labeled drawings of wave models showing wavelengths, crests, amplitudes, and the origin.

2. Draw waves of red light and blue light, showing their differences. Label the wave features.

3. Using colored pencils, sketch the emission spectrums of three different elements, showing how they might appear when viewed using a spectroscope.

4. Sketch white light passing through a prism, creating a visible spectrum. Use colored pencils.

Name ______________________________ Date ______________

Constructing the Periodic Table

Two scientists, one a Russian chemist named Dmitri Mendeleev and the other a British physicist, Henry Moseley, made significant contributions to the development of the modern Periodic Table of the Elements. Describe the work of each of these scientists below. Use the back of this page or a separate sheet of paper for your answers.

Dmitri Mendeleev

Henry Moseley

Describing the Modern Periodic Table of the Elements

One of the most common resources for the chemist is the Periodic Table of the Elements. Knowing how the table is constructed and how to understand the information available is one of the primary skills any student of chemistry must acquire.

Explain the basic structure and arrangement of the periodic table, and discuss the electron configurations and periodicity of the nobel gases, the representative elements, the transition metals, and the inner transition metals. Use the back of this page or a separate sheet of paper for your answers.

Name ______________________________ Date ______________

Going Deeper Into the Periodic Table

Some of the characteristics of the elements revealed through examination of the periodic table are less obvious, but no less significant. Explain or describe the trends revealed in the periodic table in regard to atomic size, ionization energy, electron affinity, ionic size, and electronegativity.

1. atomic size

2. ionization energy

3. electron affinity

4. ionic size

5. electronegativity

Name ______________________________ Date ______________

Looking at the Groups

The elements that make up the Periodic Table of the Elements have some common chemical and physical properties within each group, as well as some important differences. Listed below are seven of the groups. Note some of their common characteristics and features, as well as differences between elements in the group. Name some elements in each group.

1. alkali metals

2. alkaline earth metals

3. aluminum group

4. carbon group

5. nitrogen group

6. oxygen group

7. halogen group

Name ______________________________ Date ______________

Periodic Properties

Demonstrate your understanding of the trends followed in the Periodic Table of the Elements by correctly identifying whether a characteristic increases, decreases, or remains constant when you move in a certain direction across the table.

Part A

When you move from **left** to **right** across the periodic table, do the following increase, decrease, or remain constant? Write I, D, or C in the blank to the left of each property or characteristic stated.

_____ 1. electronegativity

_____ 2. shielding effect

_____ 3. nuclear charge

_____ 4. ionization energy

_____ 5. electron affinity

_____ 6. atomic radii

Part B

When you move from **top** to **bottom** down the periodic table, do the following increase, decrease, or remain constant? Answer with an I, D, or C.

_____ 7. ionization energy

_____ 8. electron affinity

_____ 9. electronegativity

_____ 10. shielding effect

_____ 11. atomic radii

_____ 12. nuclear charge

_____ 13. ionic size

Name ________________________________ Date ______________

FRANK SCHAFFER'S **CHEMISTRY FOR EVERYDAY**

Electron Configurations

Listed below are several elements, along with boxes representing their orbital filling and their electron configuration. Using the electron configurations, drop up arrows or up and down arrows showing the orbital filling for each element based upon:

1) The Aufbau principle—electrons enter orbitals of lowest energy first.
2) The Pauli exclusion principle—an atomic orbital may describe at most two electrons.
3) Hund's rule—when electrons occupy orbitals of equal energy, one electron enters each orbital until all the orbitals contain one electron with spins parallel. Second electrons then add to each orbital so that their spins are paired with the first electrons in the orbital.

	Orbital Filling						
Element	**1s**	**2s**	**2px**	**2py**	**2pz**	**3s**	**Electron configuration**
H	☐	☐	☐	☐	☐	☐	$1s^1$
He	☐	☐	☐	☐	☐	☐	$1s^2$
Li	☐	☐	☐	☐	☐	☐	$1s^2\ 2s^1$
Be	☐	☐	☐	☐	☐	☐	$1s^2\ 2s^2$
B	☐	☐	☐	☐	☐	☐	$1s^2\ 2s^2\ 2p^1$
C	☐	☐	☐	☐	☐	☐	$1s^2\ 2s^2\ 2p^2$
N	☐	☐	☐	☐	☐	☐	$1s^2\ 2s^2\ 2p^3$
O	☐	☐	☐	☐	☐	☐	$1s^2\ 2s^2\ 2p^4$
F	☐	☐	☐	☐	☐	☐	$1s^2\ 2s^2\ 2p^5$
Ne	☐	☐	☐	☐	☐	☐	$1s^2\ 2s^2\ 2p^6$
Na	☐	☐	☐	☐	☐	☐	$1s^2\ 2s^2\ 2p^6\ 3s^1$
Mg	☐	☐	☐	☐	☐	☐	$1s^2\ 2s^2\ 2p^6\ 3s^2$

Name ______________________________ Date ______________

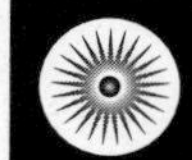

Terms of Ionic Bonding

Match the description or definition with the term by writing the letter of the term in the space provided.

Definition or Description

1. ______ Electrons in the highest occupied energy level of an element's atoms.
2. ______ A way of representing valence electrons.
3. ______ Atoms react by changing the number of their electrons so as to acquire the stable electron structure of a noble gas.
4. ______ Anions of chlorine and the other halogens.
5. ______ The forces of attraction that bind oppositely charged ions together.
6. ______ The number of ions of opposite charge that surround each ion in a crystal.
7. ______ The attraction of the free-floating valence electrons for the positively charged metal ions.
8. ______ Able to be drawn into wire.
9. ______ Ability to be hammered into different shapes.
10. ______ Solid solutions made by melting different metals together and cooling them.
11. ______ A solid metal solution containing the metal mercury.
12. ______ An alloy where the atoms of different metals may replace each other in the metal crystals.
13. ______ When the small atoms of a metal fit between the larger atoms of another metal in an alloy.
14. ______ Another name for compounds formed by ionic bonding.

A. alloy	F halide ion	K. octet rule
B. amalgam	G. interstitial alloy	L. salt
C. coordination number	H. ionic bond	M. substitutional alloy
D. ductile	I. malleable	N. valence electron
E. electron dot structure	J. metallic bond	

Name ______________________________ Date ______________

Investigating Ionic Bonds

Atoms are held together in compounds by chemical bonds. There are several types of chemical bonds including ionic bonds. Answer the following questions about ionic bonds and the properties of substances held together in this way:

1. Explain how a chemical bond takes place.

2. What are valence electrons? Why are they important to chemical bonding?

3. Why is it possible for noble gases to exist as isolated atoms?

4. Explain ionic bonds.

5. What are some of the characteristics of ionic compounds?

6. Describe a metallic bond. Explain how this type of bond is associated with the characteristics of metals.

Name ______________________________ Date ______________

Put the Metal to the Metal

Metals are a type of ionic bond formed when free-floating valence electrons are attracted to the positively charged metal ions. Metals are simple crystalline solids and can be combined in solid solutions called alloys. Alloys are formed by melting different metals together in set proportions, then cooling them.

Name twelve different alloys. Give the percentage composition of each metal involved. List some uses for each alloy.

1.

2.

3.

4.

5.

6.

7.

8.

9.

10.

11.

12.

Name ______________________________ Date ______________

Setting Examples

It is important to make connections between a chemistry concept and the rest of the world. When talking about ionic and covalent bonds, it is easy to lose that connection.

1. Diatomic elements: Name 5 examples. List their properties and uses.

   a)

   b)

   c)

   d)

   e)

2. Covalent compounds: Name 10 examples. List their properties and uses.

   a)

   b)

   c)

   d)

   e)

   f)

   g)

   h)

   i)

   j)

3. Ionic compounds: Name 5 examples. List their properties and uses.

   a)

   b)

   c)

   d)

   e)

Name ____________________ Date ____________

Bonds: Covalent Bonds

Ionic bonds are created because of electrostatic attraction between atoms that have formed positive or negative ions. There are other kinds of chemical bonds that are possible. These involve the sharing of electrons between atoms.

Explain or describe the following types of chemical bonds. Give at least one example of each.

1. single covalent bonds

2. double and triple covalent bonds

3. coordinate covalent bonds

4. polar bonds

5. nonpolar bonds

Name ______________________________ Date ______________

The Vocabulary of Covalent Bonds

Write a definition or description of the following terms related to covalent bonds:

1. antibonding orbital
2. bond dissociation energy
3. bonding orbital
4. dipole
5. dipole interaction
6. dispersion force
7. hybridization
8. hydrogen bond
9. molecular orbital
10. network solid
11. paramagnetic
12. pi bond
13. polar molecule
14. resonance
15. sigma bond
16. structural formula
17. tetrahedral angle
18. unshared pair
19. van der Waals force
20. VSEPR theory

Covalent Versus Ionic Compounds: A Comparison Chart

Complete the chart below comparing the types of bonds and compounds formed.

	Units	Type of Bond	Elements	Physical State	Melting Point	Solubility in Water	Conductivity in aqueous solution
Covalent compounds							
Ionic compounds							

Name ______________________________ Date ______________

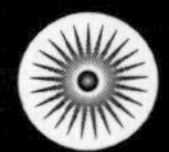

Vocabulary: Water and Solutions

Water is a simple molecule of hydrogen and oxygen, yet it is the most important compound on our planet. Water has been discovered on other worlds, too, including the Moon and Mars. Without water, life as we know it could not exist.

Match the definitions or descriptions with the letter of the terms at the bottom of the page.

1. ______ The inward pull or force that tends to minimize the surface area of a liquid.
2. ______ A wetting agent that serves to decrease the surface tension of water.
3. ______ Water samples containing dissolved substances.
4. ______ The dissolving medium in a solution.
5. ______ The dissolved particles in a solution.
6. ______ This occurs when a solute dissolves.
7. ______ The water in a crystal.
8. ______ A hydrate losing the water of hydration because the hydrate has a vapor pressure higher than that of the water vapor in air.
9. ______ A quality of compounds that removes water from the air.
10. ______ A hygroscopic substance used as a drying agent.
11. ______ A substance that becomes wet when exposed to normally moist air.
12. ______ Compounds that conduct an electric current in aqueous solutions.
13. ______ Compounds that will not conduct an electric current in aqueous solutions.
14. ______ When in solution, only a fraction of the solute exists as ions.
15. ______ When in solution, a large portion of the solute exist as ions.
16. ______ Mixtures from which some of the particles will settle slowly upon standing.
17. ______ Mixtures containing particles that are intermediate in size between those of suspensions and true solutions.
18. ______ Scattering of visible light in all directions by colloidal particles.
19. ______ The chaotic movement of colloidal particles.
20. ______ Colloidal dispersions of liquids in liquids.

A. aqueous solution	F. electrolyte	K. solvation	P. surfactant
B. Brownian motion	G. emulsion	L. solvent	Q. Tyndall effect
C. colloid	H. hygroscopic	M. strong electrolyte	R. water of hydration
D. deliquescent	I. nonelectrolyte	N. surface tension	S. weak electrolyte
E. effloresce	J. solute	O. suspension	T. desiccant

Molecules: Staying in Shape

Molecules exist in three-dimensions that are not represented by electron dot structures and structural formulas. Draw the six following common shapes of molecules.

1. Linear triatomic

2. Bent triatomic

3. Pyramidal

4. Tetrahedral

5. Trigonal planar

6. Trigonal bipyramidal

Name ______________________ Date ____________

FRANK SCHAFFER'S **CHEMISTRY FOR EVERYDAY**

Solubility of Solids at Different Temperatures

One of the factors affecting the solubility of substances in water is the temperature of the water. Table A gives some experimental data listing the concentration of four substances dissolved in water at different temperatures. Graph the data; then answer the questions below.

Table A
(Concentration g/100 g H_2O)

Temperature °C	KNO_3	Na_2SO_4	NH_4Cl	NaCl
0°	18	60	30	38
10°	20	58	34	39
20°	30	56	38	40
30°	44	54	42	41
40°	60	52	46	42
50°	82	50	56	43
60°	108	48	64	44

1. Describe the relationship between solubility and water temperature for each of the four substances.

2. Draw a line on the graph below showing the general relationship between temperature and the solubility of gases in water.

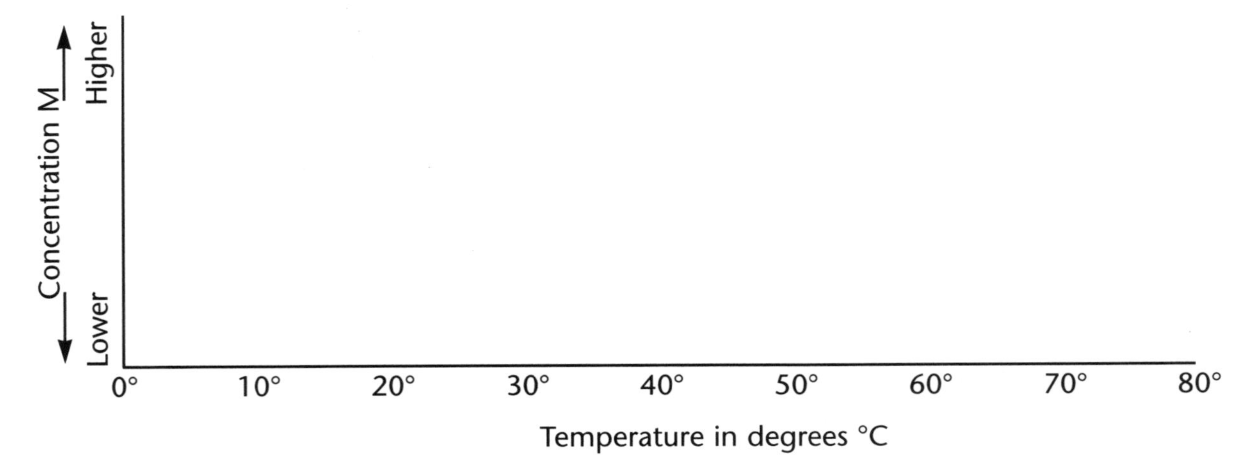

Name ______________________________ Date ______________

FRANK SCHAFFER'S CHEMISTRY FOR EVERYDAY

Solutions to Solutions

Answer the following questions related to solutions and their properties:

1. What are some of the factors affecting solubility?

2. How does chromatography work?

3. What is molarity?

4. Explain why, when diluting acid, you always add the acid to the water and not the reverse.

5. How is a percent solution prepared?

6. What are colligative properties and what do they have to do with solutions?

Name ______________________ Date ______________

Searching for Solutions

Solutions are homogeneous mixtures that exist as solids, liquids, or gases, though chemists usually work with liquid solutions.

Define or explain the following terms related to solutions and their properties:

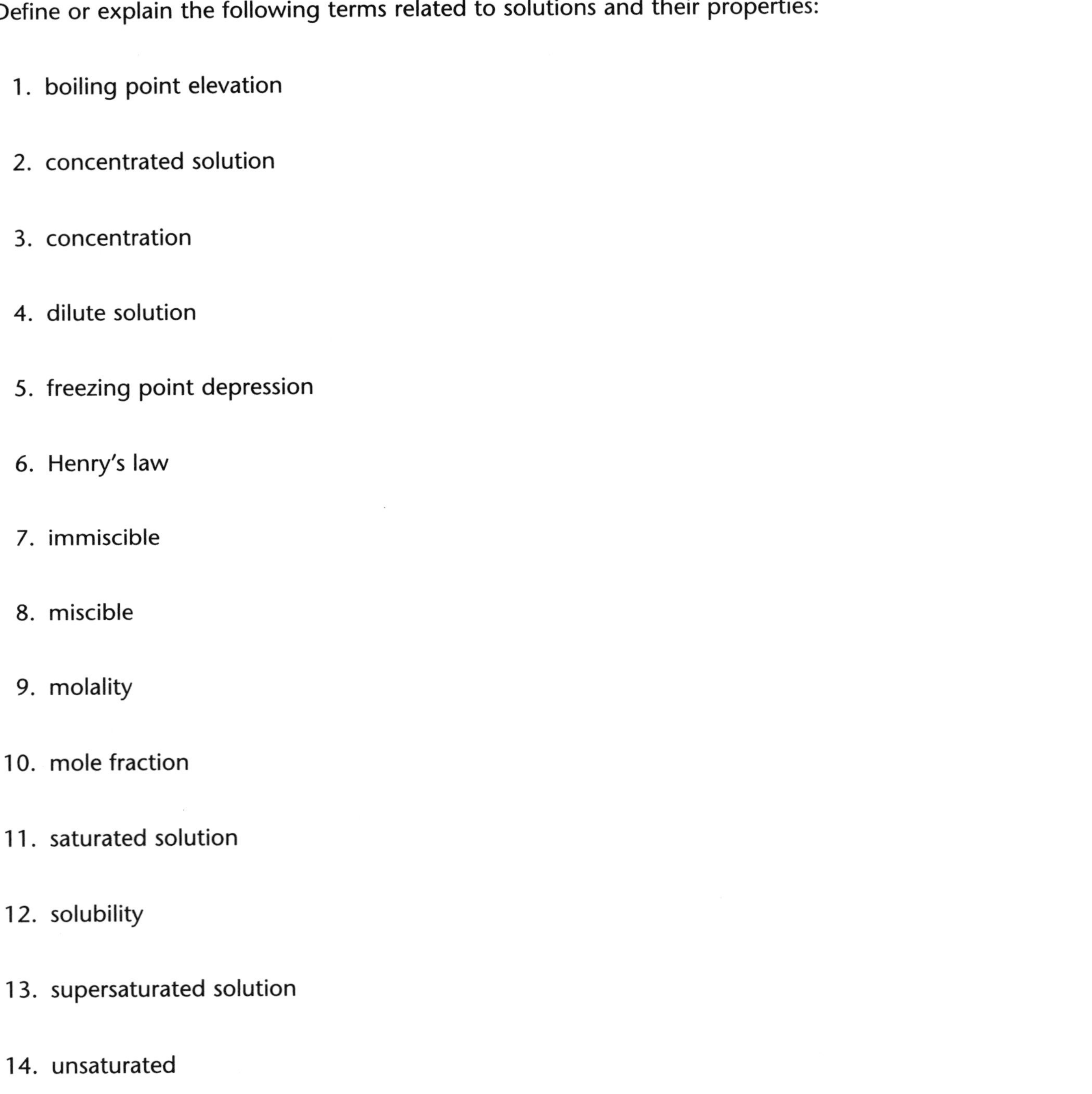

1. boiling point elevation
2. concentrated solution
3. concentration
4. dilute solution
5. freezing point depression
6. Henry's law
7. immiscible
8. miscible
9. molality
10. mole fraction
11. saturated solution
12. solubility
13. supersaturated solution
14. unsaturated

Name ______________________________ Date ____________

Matching Reaction Rates

These terms, definitions, or descriptions concern chemical reaction rates and equilibrium. Write the letter of the correct matching term in the space next to the number.

1. ______ Nonspontaneous reactions absorbing free energy
2. ______ The measure of the disorder of a system
3. ______ Said of spontaneous reactions that release free energy
4. ______ A substance that interferes with catalysis
5. ______ Reactions carried out with heterogeneous mixtures of reactants
6. ______ A substance that increases the rate of a reaction without being used up itself
7. ______ A product of a reaction that immediately becomes a reactant in another reaction
8. ______ Energy available to do work when a reaction occurs
9. ______ Things move spontaneously in the direction of maximum chaos or disorder
10. ______ The number of atoms, ions, or molecules that react in a given time to form products
11. ______ Reactions that are known to produce the written products
12. ______ Another name for the very unstable activated complex formed momentarily during a reaction

A. catalyst	D. exergonic	J. free energy	G. inhibitor
B. endergonic	E. reaction rate	K. heterogeneous reactions	H. law of disorder
C. entropy	F. spontaneous reactions	L. transition state	I. intermediate

Name ______________________________ Date ____________

FRANK SCHAFFER'S **CHEMISTRY FOR EVERYDAY**

Chemical Reaction Rates

Define the following terms related to chemical reaction rates and equilibrium:

1. activated complex
2. activation energy
3. change in Gibbs free energy
4. chemical equilibrium
5. elementary reaction
6. equilibrium constant
7. equilibrium position
8. Le Chatelier's principle
9. nonspontaneous reaction
10. rate law
11. reaction mechanism
12. reversible reaction
13. specific rate constant
14. standard entropy

Name ______________________ Date ____________

Explaining Chemical Reaction Rates

Answer the following questions in complete sentences:

1. Explain what the activation energy barrier is, and why it must be overcome for a reaction to take place.

2. What is the collision theory of reactions?

3. What is entropy, and what does it have to do with reactions?

4. What factors affect the rate of a chemical reaction?

5. What is a catalyst, and how do they work?

6. Under what circumstances are reactions reversible or irreversible?

Name ______________________ Date ______________

The Acid-Base Test

Two of the most widely used categories of chemicals in industry and in households everywhere are acids and bases. Answer the questions below concerning the detection and measurement of these important compounds.

1. Explain the three theories often used to classify substances as acids or bases. These theories were proposed by Arrhenius, Brönsted and Lowry, and Lewis.

2. How is the strength of an acid or base determined?

3. What is the pH scale, and how is pH measured?

Name ______________________________ Date ____________

Acids and Bases

Define the following terms as they relate to acids and bases:

1. acid dissociation constant
2. acidic solution
3. alkaline solution
4. amphoteric
5. base dissociation constant
6. basic solution
7. conjugate acid
8. conjugate base
9. conjugate acid-base pair
10. diprotic acid
11. hydrogen-ion acceptor
12. hydronium ion
13. hydroxide ion
14. ion-product constant for water
15. monoprotic acid
16. neutral solution
17. self-ionization
18. strong acid
19. strong base
20. triprotic acid
21. weak acid
22. weak base

Name ______________________________ Date ______________

FRANK SCHAFFER'S **CHEMISTRY FOR EVERYDAY**

Going Deeper: Acids and Bases in Your Life

Acids and bases are present in our bodies and throughout our environment, as well as being the most common industrial chemicals. What are some of the uses for the acids and bases listed below?

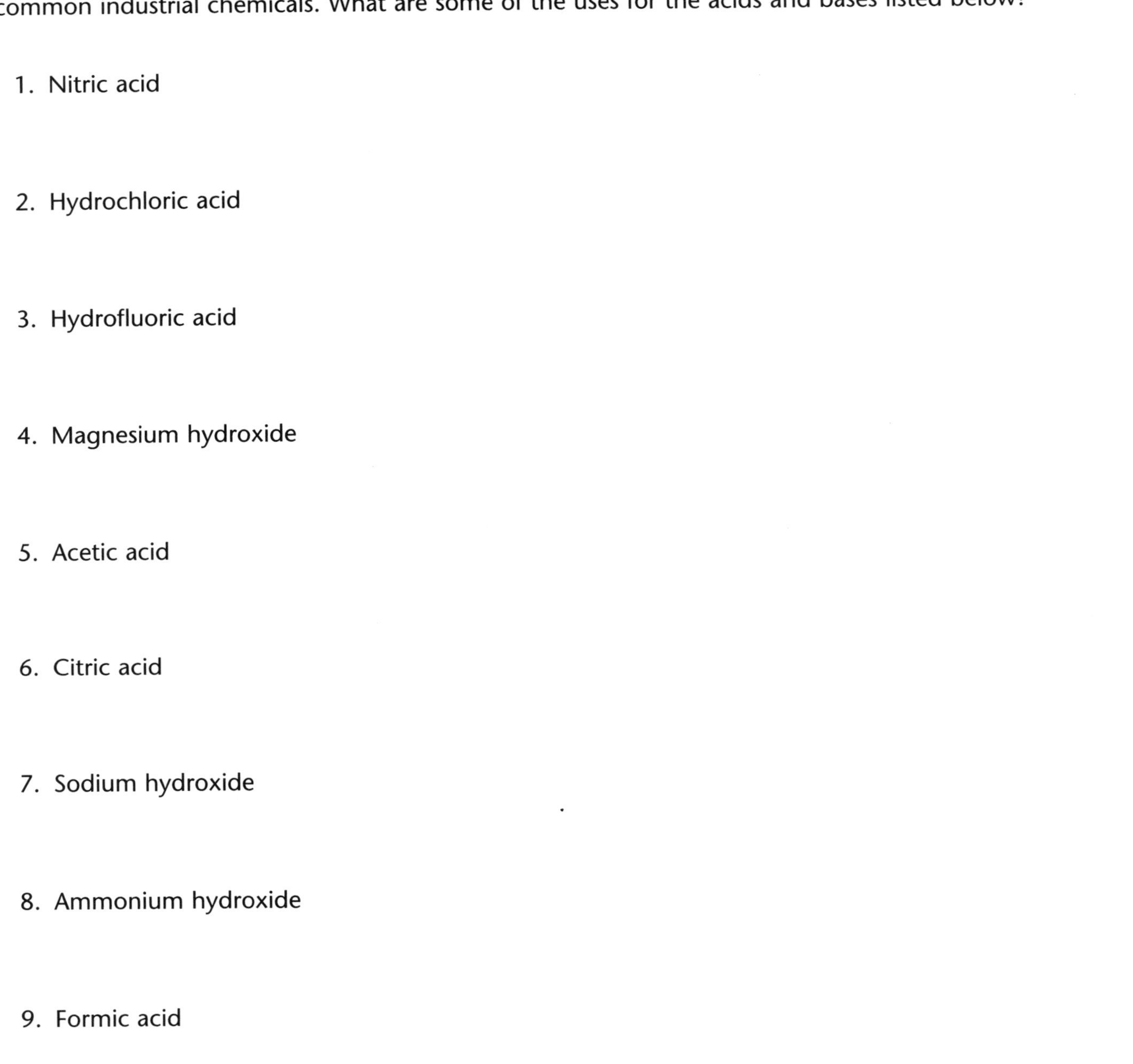

1. Nitric acid

2. Hydrochloric acid

3. Hydrofluoric acid

4. Magnesium hydroxide

5. Acetic acid

6. Citric acid

7. Sodium hydroxide

8. Ammonium hydroxide

9. Formic acid

Name ______________________________ Date ______________

Oxidation-Reduction

These definitions and descriptions are related to oxidation-reduction reaction. Write the letter of the correct matching term in the space next to the number.

1. ______ The chemical changes that occur when electrons are transferred between reactants.

2. ______ Another name for oxygen-reduction reactions.

3. ______ The loss of oxygen from the compound.

4. ______ The substance in a redox reaction that donates electrons.

5. ______ The substance in a redox reaction that accepts electrons.

6. ______ A positive or negative number assigned to an atom according to a set of arbitrary rules.

7. ______ A redox equation is balanced by comparing the increases and decreases in oxidation numbers.

8. ______ Balancing redox equations by balancing the oxidation and reduction half-reactions.

9. ______ An equation showing either the reduction or the oxidation of a species in an oxidation-reduction reaction.

10. ______ These do not change oxidation number or composition during a reaction.

11. ______ The combination of an element with oxygen to produce oxides.

A. half reaction	G. oxidizing agent
B. half-reaction method	H. redox reaction
C. oxidation	I. reduction
D. oxidation number	J. spectator ion
E. oxidation number change method	K. reducing agent
F. oxidation-reduction reaction	

Name ______________________ Date ______________

Comparing Oxidation and Reduction

Listed below are events or results of oxidation-reduction reactions under the heading PROCESSES. In the columns to the right, tell what happens during OXIDATION and REDUCTION concerning these processes.

PROCESSES	OXIDATION	REDUCTION
1. Number of electrons		
2. Where electrons go		
3. Status of oxygen		
4. Status of hydrogen		
5. Oxidation number		

Give three examples of balanced oxidation reactions and three examples of reduction reactions.

6.

7.

8.

State three rules for assigning oxidation numbers.

9.

10.

11.

Name ______________________________ Date ______________

Chemistry in Your Life

The chemical reactions of oxidation-reduction are present everywhere in your life, not just in chemistry class. Burning wood in a fireplace or gasoline in an automobile, or the conversion of food into energy for our bodies are all examples of oxidation reactions.

Answer the following questions related to the chemistry of our daily lives:

1. Rust, the corrosion of certain substances, is a form of oxidation. What is rust? How does it affect our lives? What conditions promote corrosion? What is being done to counteract its effects?

2. What properties do substances like aluminum and chromium have that makes them important in automobile construction?

3. What special problems do the metal hulls of boats and ships experience with corrosion? What can be done to combat these problems?

4. Carbon monoxide is a by-product of automobile exhaust. What chemical properties does CO have that make it a danger to humans and other animals?

5. Sulfur dioxide is a common food additive. Why?

Name ______________________________ Date ____________

FRANK SCHAFFER'S **CHEMISTRY FOR EVERYDAY**

Generating Electricity With Chemical Energy

Electrical energy can be generated with spontaneous redox reactions. Electrical energy can be used to cause nonspontaneous chemical reactions in what are termed electrochemical cells. Define the terms related to electrochemistry.

1. anode
2. battery
3. cathode
4. cell potential
5. dry cell
6. electrical potential
7. electrochemical cell
8. electrochemical process
9. electrode
10. electrolysis
11. electrolytic cell
12. fuel cell
13. half-cell
14. reduction potential
15. salt bridge
16. standard cell potential
17. standard hydrogen electrode
18. voltaic cell

Name ______________________________ Date ______________

FRANK SCHAFFER'S **CHEMISTRY FOR EVERYDAY**

Electrochemical Processes

Voltaic cells are common sources of electrochemical energy. The types in common use today include dry cells, lead storage batteries, and fuel cells. In industry, the process is often reversed during the manufacture of many substances in what is called electrolysis.

Answer the following questions concerning electrochemical processes:

1. Describe dry cells. Explain how they work. List some of their uses.

2. What are lead storage batteries? How do they work? What are some of their applications?

3. How are fuel cells different from other voltaic cells. What are their applications?

4. Explain how electrolysis is used to obtain supplies of an important substance.

Understanding the Voltaic Cell

Explain how a voltaic cell works. Include a labeled diagram or drawing as part of your explanation. Use a separate sheet of paper for your answers.

Name ______________________ Date ____________

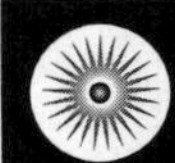

FRANK SCHAFFER'S **CHEMISTRY FOR EVERYDAY**

Chemistry, Sources, and Uses of Metals

Around 75% of the elements on the periodic table are classified as metals. Most have common physical and chemical characteristics within their respective groups. Metals also have unique properties and characteristics that make them important to us. Describe the chemistry, distinctive properties, and uses of the following three groups of metals:

1. alkali metals

2. alkaline earth metals

3. transition metals

Name ______________________ Date ____________

FRANK SCHAFFER'S CHEMISTRY FOR EVERYDAY

Really Heavy Metals

Several metals are important as essential minerals in the human body. Zinc, iron, and magnesium are three of them. Some of those listed below are also important for humans in very small amounts, but in larger amounts are toxic. The rest are toxic in any amount and may be fatal. Explain the effects these metals have on humans or living systems.

1 mercury

2. lead

3. cadmium

4. arsenic

5. copper

Answer Key

READING THE PERIODIC TABLE OF THE ELEMENTS, page 1

1. a) hydrogen, b) aluminum, c) iron, d) sodium, e) europium, f) lawrencium, g) cesium, h) krypton, i) platinum, j) xenon
2. a) He, b) I, c) Pu, d) Au, e) Pb, f) Sn, g) U, h) Es, i) C, j) Cl
3. a) 1, b) 13, c) 26, d) 11, e) 63, f) 103, g) 55, h) 36, i) 78, j) 54
4. a) 4, b) 126.9, c) 244, d) 197, e) 207.2, f) 118.7, g) 238, h) 254, i) 12, j) 35.5
5. lithium, sodium, potassium, rubidium, cesium, francium
6. gold, silver
7. neon, argon, krypton, xenon
8. The elements are listed based upon their atomic numbers.

PERIODIC TABLE CROSSWORD, page 2

ACROSS	DOWN
4. IRIDIUM	1. SILICON
5. TERBIUM	2. BISMUTH
7. WOLFRAM	3. CARBON
8. ARGON	6. FRANCIUM
9. HYDROGEN	10. COPPER
11. IRON	12. RADON
13. BARIUM	14. CESIUM
17. RUTHENIUM	15. PROMETHIUM
18. TECHNETIUM	16. NEON
20. CHROMIUM	19. GOLD
21. LAWRENCIUM	

SCIENTIFIC METHOD APPLIED TO CHEMISTRY, page 3

1. The question being asked, or the problem to be solved.
2. Gathering information about the objective, including the examination of previous related experiments.
3. A proposed explanation related to the question that includes a prediction or educated guess . . . if this is true, then this will happen . . .
4. The hypothesis will suggest a test or experiment to test its validity.
5. What happens during the experiment that the observer is able to detect.
6. The information, numbers, values, and other quantifiable measurements made during the experiment.
7. The organization of the data into tables, charts, descriptive statements.
8. A decision as to whether the hypothesis or prediction was correct or proven, based upon the results.
9. The experiment can be repeated and confirmed because of similar results, or repeated with modifications based upon a new prediction.

DETERMINING PHYSICAL PROPERTIES, page 4

1.	Liquid	0°C	100°C	Clear, tasteless
2.	Solid	1535°C	2750°C	Magnetic
3.	Gas	−210°C	−195.8°C	Colorless
4.	Gas	−218°C	−183°C	Colorless
5.	Solid	1083°C	2595°C	Good conductor
6.	Solid	113°C	445°C	Yellow, sharp odor
7.	Solid	3727°C	4830°C	Carbon and diamond are forms
8.	Gas	−101°C	−34.7°C	Yellow-green color
9.	Solid	1063°C	2970°C	Very dense
10.	Solid	1769°C	4530°C	Silver-white color
11.	Solid	960.8°C	2210°C	Silver-white color

12. Answers will vary: Color, solubility, mass, odor, hardness, density, electrical conductivity, magnetism

STATE PROPERTIES, page 5

1. definite; definite; definite
2. definite or fixed; variable (shape of container); variable (shape of container)
3. definite or fixed; definite or fixed; indefinite or variable
4. slight expansion; medium or more expansion; large or great expansion
5. very slight; very slight; yes
6. vibrate in place; slow moving; constant, chaotic, rapid movement
7. ice, sugar, salt; water, blood, milk, mercury; oxygen, nitrogen, helium
8.

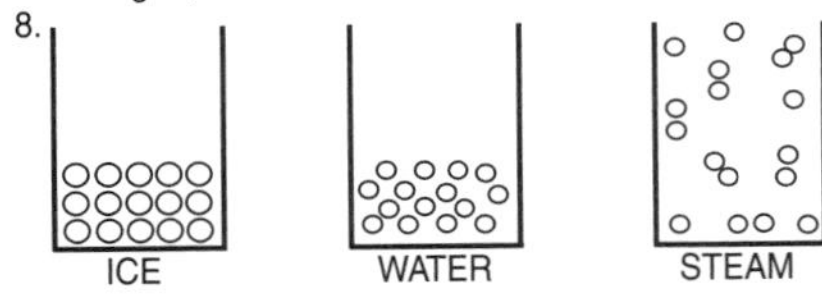

UNDERSTANDING MATTER AND ENERGY, page 6

1. A chemical change occurs when there is a change in the composition of a substance as a result of a chemical reaction. Example: When carbon, oxygen, and hydrogen in wood are burned, they change into charcoal, water, and energy.
2. Chemical property is the ability of a substance to undergo certain chemical reactions to produce new substances. Example: When iron, water, and oxygen combine to form iron oxide. The ability of iron to rust is a chemical property of iron.
3. A chemical reaction occurs when one or more substances undergo a chemical reaction, changing into new substances. Example: When gasoline burns, it produces water, carbon dioxide, and other gases, along with heat energy.
4. Often the first letter or two of an element's name is used as a shorthand symbol for that element in chemical reactions. Example: Oxygen's chemical symbol is O and the chemical symbol for helium is He.
5. Kinetic energy is energy of motion. Example: The energy in a flywheel or a thrown baseball.
6. In any chemical or physical process, energy is neither created nor destroyed. It may be converted from one form to another. Example: When boiling water produces steam and turns a turbine, heat energy is converted to mechanical energy.
7. In any chemical or physical reaction, mass is neither created nor destroyed. The mass of the products is equal to the mass of the reactants. Example: If a piece of wood is burned in a chamber, the ash, water vapor, and other gases will have an equal mass.
8. Physical change occurs when a substance is altered without changing its composition. Example: When a rock is ground into dust, or a wire bent, or water freezes into ice.
9. Potential energy is stored energy, the energy of position. Example: Stored fat in an animal.
10. A scientific law is a statement that summarizes the results of a broad variety of observations and experiments that describe a natural phenomenon without explaining it. Example: Newton's law states that for every action, there will be an equal but opposite reaction.

BASIC TERMS IN CHEMISTRY, page 7

ACROSS	DOWN
6. OBSERVATION	1. CHEMISTRY
9. PHASE	2. HOMOGENEOUS
10. PRODUCTS	3. MASS
11. SOLUTION	4. COMPOUNDS
12. HYPOTHESIS	5. MIXTURE
13. ENERGY	7. EXPERIMENT
16. ELEMENTS	8. SOLID
20. HEAT	11. SUBSTANCE
22. REACTANTS	14. GAS
	15. PROPERTY
	17. LIQUID
	18. MATTER
	19. THEORY
	21. DATA

RECOGNIZING PHYSICAL AND CHEMICAL CHANGES, page 8

1. A physical change is a change in a substance that does not result in the substance itself being changed. The change is reversible. If a container of water is frozen into ice, the water has been changed, but only physically. When the ice melts the water will still be there. Possible answers:
   a. water being boiled into steam
   b. sugar dissolved in water: when the water is boiled out the sugar will still be there
   c. iron filings mixed with sand
   d. a piece of copper wire being bent
   e. gold melted, then poured into a mold
   f. a chunk of coal ground into powder
   g. a tree cut down and chopped into firewood
   h. a rock cracks when water freezes in a crevice and expands
   i. oil being pumped out of the ground
   j. a rubber band being stretched until it breaks
2. A chemical change occurs when a substance is changed into another substance because of a chemical reaction and the reactants and products are different from each other. This type of change is rarely reversible. Possible answers:
   a. iron and sulfur mixed together and heated to form iron sulfide
   b. the metal sodium chemically bonded to the gas chlorine, forming salt
   c. burning coal, to form carbon, carbon dioxide, water, and other gases
   d. an orange rotting in a locker
   e. a steel nail rusting
   f. mixing flour, sugar, salt, milk, yeast, and shortening and then cooking it to make bread
   g. painting a wall (the paint oxidizes and dries)
   h. a seed develops into a mature plant, producing fruit
   i. cooking an egg
   j. when solid mercury sulfide is heated with oxygen, liquid mercury metal and sulfur dioxide gas are produced

SYSTEMS OF MEASUREMENT, page 9

1. Meter, m; **2.** Kilogram, kg; **3.** Second, s; **4.** Ampere, A; **5.** Kelvin, K; **6.** Mole, mol; **7.** Candela, cd; **8.** Cubic meter, M^3; **9.** Gram, g; **10.** Grams per cubic centimeter, g/cm^3; **11.** Degree celsius, °C; **12.** Second, s; **13.** Atmosphere, atm: Pascal, pa: Millimeters of mercury, mm hg; **14.** Jovie, J: Calorie, cal

MEASURING UP, page 10

1. Quantitative measurements are stated in specific units, usually with numbers. An example would be the position of a pencil 92.5 centimeters from your hand. These values are objective, and may be understood precisely by others needing the information. Some standard reference, such as an SI unit, is usually stated. Qualitative measurements give their results in a descriptive form usually without numbers. The above pencil may be described as being about an arms-length away. If the room feels hot or cold, the notion of temperature may be very subjective, and is of little value to the experimenter. The original length of one foot being the length of the monarch's foot is a good example of a more qualitative than quantitative measurement.
2. Lavoisier proved through very accurate measurements while conducting experiments that the law of the conservation of matter was valid when a substance was burned. He disproved the existence of phlogiston, an ingredient said to be in all burnable substances. Previously people believed that when a substance was burned, the phlogiston escaped, leaving only ash. He proved that something cannot burn without the presence of oxygen.
3. Accuracy is how close a measurement comes to the actual value or dimension of the object measured. If an object with an actual mass of 200.19 g is measured and the measurement is 200.185 g, the accuracy is quite close. Precision is dealing with the issue of consistency of measurements. If a device always gives the same or close to the same value for a measured dimension, the instrument is said to be a precision instrument. A scale in a butcher shop that gives different values for the same amount of hamburger when the hamburger is placed on and then off the scale a few times is not showing precision. A target rifle that allows consistent bulls eyes is both accurate and precise. If it always shoots high and to the left of the target, but in a very tight grouping of holes, it is precise but not accurate. If the target rifle makes holes all over the target, it is neither accurate nor precise.
4. It is a numerical value where all the digits are accurately known, plus a digit that is estimated. The measured temperature of 87.5°C, for example, has three significant figures. The 8 and the 7 are accurately known, and the .5 is an estimate, and is uncertain. If the thermometer in the above example allowed a reading of 87.55 degrees C, there would be four significant figures, with the .05 value being the uncertain estimate. Sometimes the presence of zeros in numbers can be misleading, so there are established rules for determining significant figures.
5. An answer cannot be more precise than the least precise measurement, so all the numbers are rounded off to that value. Numbers with more than that number of significant figures are rounded off. If the digit immediately following the last significant digit is less than five, all the digits after the last significant place are dropped. If the digit is five or more, the value of the digit in the last significant place is increased by one.

MEASURING METRICALLY WITH METERS, page 11

1. kilometer, km, 1km = 1000 m
2. meter, m, 1m = 100 cm
3. decimeter, dm, 10 dm = 1m
4. centimeter, cm, 100 cm = 1m
5. millimeter, mm, 1,000 mm = 1m
6. micrometer, μm, 1,000,000 μm = 1m
7. nanometer, nm, 1,000,000,000 nm = 1m
8. kilo, k, 1,000 times larger than preceding unit
9. deci, d, 10 times smaller than unit it precedes
10. centi, c, 100 times smaller than the unit it precedes
11. milli, m, 1,000 times smaller than the unit it precedes
12. micro, μ, 1,000,000 times smaller than the unit it precedes
13. nano, n, 1,000 million times smaller than the unit it precedes

SPECIFIC GRAVITY SPECIFICS, page 12

1.	AL	2.70 g/cm³	2.94
2.	Be	1.85	2.02
3.	C	2.26	2.46
4.	Fe	7.86	8.57
5.	Na	0.97	1.06
6.	Ti	4.51	4.92
7.	Cr	7.19	7.84
8.	Cu	8.96	9.8
9.	Ag	10.5	11.45
10.	Au	19.3	21.05
11.	Ba	3.5	3.82
12.	U	19.07	20.8

PROBLEMS WITH NUMBERS?, page 13

1. (3), **2.** (2), **3.** (4), **4.** (4), **5.** (3), **6.** (5), **7.** (4), **8.** (3), **9.** (3), **10.** (5), **11.** (88.5), **12.** (.000863), **13.** (67.0), **14.** (8.50×10^3), **15.** (12.2), **16.** (1.2×10^3), **17.** (2.5×10^4), **18.** (8.1×10^7), **19.** (7×10^{-3}), **20.** (3.5×10^{-4}), **21.** (300,000), **22.** (230), **23.** (.006), **24.** (.2700), **25.** (6,100,000)

DETERMINING SPECIFIC GRAVITY, page 14

Answers will vary depending on the accuracy of measurements and minerals selected. Specific gravities for a few commonly available minerals are:

mica - 2.75	chalcopyrite - 4.1 to 4.3
pyrite - 4.8 to 5.2	hematite - 4.9 to 5.3
magnetite - 4.2 to 4.4	sphalerite - 3.9 to 4.1
quartz - 2.6	calcite - 2.7
obsidian - 2.2 to 2.8	fluorite - 3 to 3.25
apatite - 3.2	microcline - 2.55

ATOMIC STRUCTURE, page 15

1. G, **2.** L, **3.** A, **4.** K, **5.** F, **6.** D, **7.** E, **8.** J, **9.** I, **10.** C, **11.** B, **12.** H

USING THE PERIODIC TABLE OF THE ELEMENTS, page 16

1.	hydrogen	1	1	1	1	0
2.	helium	2	4	2	2	2
3.	carbon	6	12	6	6	6
4.	nitrogen	7	14	7	7	7
5.	oxygen	8	16	8	8	8
6.	cobalt	27	59	27	27	32
7.	copper	29	64	29	29	35
8.	iron	26	56	26	26	30
9.	uranium	92	238	92	92	146
10.	krypton	36	84	36	36	48

PROBLEMS WITH THE ATOMIC STRUCTURE, page 17

1. This number represents the weighted average of the masses of all the isotopes of that element and reflects not only the different masses included, but also the relative numbers of each isotope.
2. Rutherford found out in 1911 that the mass of the atom is concentrated in a small area (the nucleus) and that the rest of the atom is mostly empty space. The nucleus is very dense and the rest of the atom has a much lower density.
3. Hydrogen has three isotopes, one with no neutron, one with one, and one with two neutrons. The isotope with one neutron, deuterium, is twice as heavy as the more common form, with no neutrons. Tritium, with two neutrons, is three times the mass as the more common form. Over 99% of the natural hydrogen has no neutrons. Only about .015% of the hydrogen atoms have one neutron. Those with two neutrons are barely measurable. Tritium is unstable and will slowly disintegrate.
4. He assumed that gases were tightly packed together. We know that is not true. He assumed that the simplest compound between two elements required one atom of each. That is incorrect. He stated that the atom is indivisible and that was incorrect too.

5. A glass tube containing two metal disks, a cathode, and an anode, is filled with a low pressure gas. When connected to a source of high-voltage, the tube glows because electrons travel from the cathode, or negative metal disk, to the anode, the positive disk. This glowing beam is called a cathode ray. Because the ray is composed of negative particles, a magnet or electrically charged (positive) plate will deflect it. Television picture tubes and computer monitors are specialized cathode ray tubes.

Cathode Ray Tube

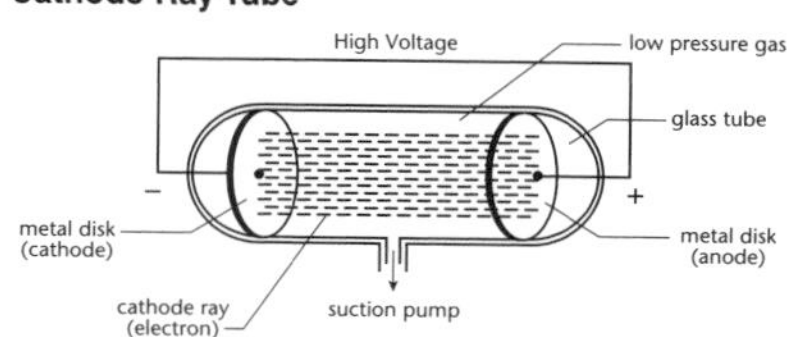

EXPLAINING CHEMICAL TERMS, page 18

1. Compounds composed of only two elements.
2. Shows the kinds and numbers of atoms in the smallest representative unit of the substance.
3. The lowest whole-number ratio of ions in an ionic compound.
4. Compounds composed of positive and negative ions.
5. In any chemical compound the elements are always combined in the same proportion by mass.
6. Whenever two elements form more than one compound, the different masses of one element that combine with the same mass of the other element are in the ration of small whole numbers
7. Compounds that are composed of molecules. Most are composed of two or more nonmetallic elements.
8. Shows the numbers and kinds of atoms present in a molecule of a compound.
9. A chart arranging the elements in rows and columns based upon similarities in their properties.
10. Tightly bound groups of atoms that behave as a unit and carry a charge.
11. The Group A elements in the periodic table, which illustrate the entire range of chemical properties.
12. Contains atoms of three different compounds.
13. The B group elements.

FORMULAS AND FORMULA NAMES, page 19

1. Ba SO_4, 2. $Al(HCO_3)$, 3. Na C10, 4. $Pb(CrO_4)_2$, 5. $Al(OH)_3$, 6. $KC10_2$, 7. $Zn(H_2SO_4)_2$, 8. CS_2, 9. $HCLO_4$ 10. aluminum iodide, 11. iron (II) oxide or ferrous oxide, 12. copper (I) sulfate or cuprous sulfide, 13. calcium selenide, 14. magnesium phosphate, 15. lithium chromate, 16. potassium silicate, 17. lithium flouride

MODELS OF MOLECULAR COMPOUNDS, page 20

1. 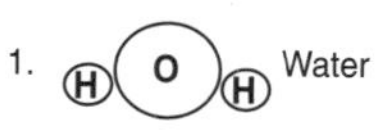

2. O C Carbon monoxide

3. O C O Carbon dioxide

4. H C C H (with H above and below each C) Ethene

5. H C C O H (with H above and below each C) Ethyl alcohol

6. H C O H (with O above C) Formic acid

NAMES AND FORMULAS, page 21

1. For many elements in Group A, the metals in Groups 1A, 2A, and 3A lose electrons and become positive ions. The charge is equal to the group number 1, 2, or 3. Those in Groups 5, 6, and 7 form negative ions, and in 5, 6, and 7, respectively, form ions with charges of –3, –2, and –1. Those in 4A and 0 normally don't form ions.
2. The old system used a root word and suffixes. The classic name of the element is the root word and an -ous ending is used for the name of the ion with the lower of the two ionic charges. An ending of -ic is used with the higher of the two ionic charges. The disadvantage of the system is that the name doesn't tell you the charge on the ion, only the size of the charge. Examples are ferrous or ferric ions (iron) and cuprous and cupric ions (copper).
3. The Stock system uses the modern name and a Roman numeral to indicate the size of the charge of the ion, as with iron (II) ion or iron (III) ion.
4. Most end with -ite or -ate, but there are a few exceptions, including two that end with -ide: cyanide and hydroxide ions.
5. In the formula for an ionic compound, the cation (the element with the positive charge) is always named first. In monatomic cations, it's only the name of the element followed by ion. In monatomic anions, the name ends in -ide.
6. When a cation has more than one common ionic charge, the Roman numeral indicates its strength. If copper combines with oxygen at a 1:1 ratio, the copper must be Cu^{2+}; copper(II) oxide. The oxide ion is always a –2 charge. If you have copper(I) oxide, the formula is Cu_2O, and the copper ion is Cu^{1+}.

TERMS RELATED TO CHEMICAL QUANTITIES, page 22

1. Equal to a mole of a substance, this number represents 6.02 x 10 representative particles of that substance and was named after Amedeo Avogadro di Quarenga, a nineteenth century Italian scientist and lawyer, whose work made the calculation of this number possible.
2. Gives the lowest whole-number ratio of the elements in a compound.
3. The number of grams of an element that is numerically equal to the atomic mass in amu (hydrogen = 1.0 g, carbon = 12.0 g).
4. The mass of one mole of an ionic compound, equal to the formula mass in grams.
5. The mass of one mole of a compound, expressed in grams.
6. Term often used instead of gram formula mass to refer to the mass of a mole of any element or compound
7. At STP, one mole of any gas occupies a volume of 22.4 L. This quantity is the molar volume of a gas.
8. The number of representative particles of a substance, equal to Avogadro's number.
9. The percent by mass of each element in a compound.
10. The smallest particle that a particular substance commonly exists, whether it be as atoms, ions, or molecules.
11. Conditions where the temperature is zero degrees celsius and the pressure is one atmosphere.

KEY TERMS USED IN STUDYING CHEMICAL REACTIONS, page 23

1. A list of the metals in order of decreasing activity that will replace another in a compound. Example: In the list of metals lithium, potassium, barium, and those after it. Calcium will not replace any of the other metals, being listed last.
2. In this form of equation, each side has the same number of atoms of each element, obeying the law of conservation of mass.
   Example: $C+O_2 \longrightarrow CO_2$.
3. A substance that speeds up a reaction without itself being used up.
   Example: $H_2O_2 \xrightarrow{MnO_2} H_2O + O_2$. In this case, the manganese dioxide is the catalyst and is written above the arrow.
4. The formulas of the reactants on the left are connected by an arrow with the formulas of the products on the right.
   Example: $FeO + C \xrightarrow{\Delta} Fe + CO$.
5. A small whole number that appears in front of a formula in an equation, used to balance the equation. Example: In the reaction $2H_2 + O_2 \longrightarrow 2H_2O$ the number 2 in front of the hydrogen is the coefficient.
6. In this type of reaction, two or more substances react to form a single substance.
   Example: $S + O_2 \longrightarrow SO_2$.
7. Oxygen reacts with another substance, often producing energy in the form of light or heat. Example: $CH_4 + 2O_2 \longrightarrow CO_2 + 2H_2O$ (+heat, light).
8. A single compound is broken down into two or more simpler products.
   Example: $CaCO_3 \longrightarrow CaO + CO_2$.
9. Involves an exchange of positive ions between two compounds.
   Example: $AgNO_3 + KCl \longrightarrow AgCl + KNO_3$.
10. Atoms of an element replace the atoms of a second element in a compound.
    Example: $Fe + CuSO_4 \longrightarrow FeSO_4 + Cu$.
11. A chemical equation that does not indicate the relative amounts of the reactants and products.
    Example: $H_2 + N_2 \longrightarrow NH_3$.

UNDERSTANDING CHEMICAL SYMBOLS, page 24

1.$\downarrow$, 2,$\uparrow$, 3. $\xrightarrow{Pt}$, 4. $=$, 5. (S) 6. (L), 7. (ag), 8. (g), 9. +, 10.$\rightarrow$ 11. $\rightleftharpoons$, 12. $\xrightarrow{\Delta}$ or $\xrightarrow{heat}$

"FLESHING OUT" SKELETON EQUATIONS, page 25

1. Examples will vary. Example: $Fe + O_2 \longrightarrow Fe_2O_3$
2. Examples will vary. Example: Iron reacts with oxygen to form iron oxide.
3. Examples will vary. Example: $4Fe + 3O_2 \longrightarrow Fe_2O_3$

BALANCING CHEMICAL EQUATIONS, page 26

1. $2Na + Cl_2 \longrightarrow 2NaCl$
2. $2Ag + S \longrightarrow 2Ag_2S$
3. $2Mg + O_2 \longrightarrow 2MgO$
4. $2Al_2O_3 \longrightarrow 4Al + 3O_2$
5. $3H_2 + N_2 \longrightarrow 2NH_3$
6. $3C + Fe_2O_3 \longrightarrow 2Fe + 3CO$
7. $2KCl + F_2 \longrightarrow 2KF + Cl_2$
8. $2NaBr + Ag_2SO_4 \longrightarrow Na_2SO_4 + 2AgBr$
9. $Fe_2O_3 + 6HCl \longrightarrow 2FeCl_3 + 3H_2O$
10. $2Al + 3CuSO_4 \longrightarrow Al_2(SO_4)_3 + 3Cu$
11. $4P + 5O_2 \longrightarrow 2P_2O_5$
12. $2Al + 3H_2SO_4 \longrightarrow Al_2(SO_4)_3 + 3H_2$
13. $SrBr_2 + (NH_4)_2CO_3 \longrightarrow SrCO_3 + 2NH_4Br$
14. $2C_2H_6 + 7O_2 \longrightarrow 4CO_2 + 6H_2O$
15. $3Fe + 2O_2 \longrightarrow Fe_3O_4$

STOICHIOMETRY: CALCULATING QUANTITIES IN CHEMISTRY, page 27

1. The amount of product formed when a reaction is carried out in the laboratory.
2. A reaction where energy is absorbed.
3. The amount of heat that a given substance has at a given temperature and pressure.
4. A quantity more than enough to react with a limiting agent.
5. Reactions that release energy in the form of heat.
6. The change of enthalpy in the special case where one mole of a substance is completely burned.
7. The heat that is released or absorbed during a chemical reaction.
8. The substance that is used up in a reaction, limiting the amount of product that can be formed.
9. The ratio of the actual yield to the theoretical yield.
10. The change in enthalpy for a reaction in which 1 mol of a compound is formed from its elements.
11. The value obtained from the calculation of the amount of product formed during a chemical reaction.
12. An equation that includes the amount of heat produced or absorbed by a reaction.

KINETIC THEORY AND THE NATURE OF GASES, page 28

Gases are composed of particles, either as individual atoms or molecules with a very small volume and far from each other. There is a lot of empty space between particles and the particles neither repel nor attract one another. The particles move randomly and rapidly, and change directions if they collide with each other or other objects. When these collisions occur, energy is transferred from one particle to another, but the total kinetic energy remains constant.

Gas particles have kinetic energy because they are constantly in motion and can absorb kinetic energy when heated. Some of this heat increases the energy within the particles and some increases their motion. The average kinetic energy of gas particles is proportional to the temperature of the particles. If all the particles have the same temperature, they all have the same kinetic energy. There is no theoretical high limit to temperature, but there is one to low temperature, that of absolute zero, about -273 degrees C. At this temperature, particles have no motion or kinetic energy.

The collisions between gas particles and objects creates pressure, the most common being atmospheric pressure. Atmospheric pressure decreases with altitude because the number of particles are less, and therefore there are less collisions taking place. Barometers are used to measure atmospheric pressure.

Avogadro's hypothesis states that whenever we have equal volumes of gases at the same temperature and pressure, the volumes must contain equal numbers of particles. This makes sense because there is so much space between particles of any gas.

STATES OF MATTER: COMING TO TERMS, page 29

1. P, **2.** D, **3.** O, **4.** B, **5.** J, **6.** I, **7.** C, **8.** N, **9.** A, **10.** M, **11.** K, **12.** L, **13.** F, **14.** G, **15.** H, **16.** E

EXPLAINING LIQUIDS, SOLIDS, AND PLASMAS, page 30

1. Liquids are in motion, though not as active as gases. Particles of liquid are free to slide past one another, so, like gases, they flow and take the shape of their containers. The particles of liquids are held together by weak attractive forces, and most do not have enough kinetic energy to escape. The forces also reduce the amount of space between particles and liquids are much denser than gases. Liquids are a condensed state of matter, and are resistant to pressure. The interplay between the motion of particles and the attractive forces between them result in vapor pressure, the heat of vaporization, and the boiling point of a liquid.

   Vapor pressure is a result of collision between vaporized particles of the liquid colliding with the walls of a sealed container. The heat of vaporization is the conversion of a liquid to a gas or vapor below its boiling point. The boiling point is the temperature at which the vapor pressure of the liquid is equal to the external pressure.
2. In solids, the particles are packed against each other in a organized fashion and can only vibrate and rotate about fixed points. Solids are dense and incompressible and cannot flow like liquids or gases. When heated, particles vibrate more rapidly as the kinetic energy increases. At the melting point, their vibrations are strong enough to overcome the forces holding them in place and the solid begins to melt into a liquid. The freezing and melting points of a solid are the same temperature. Most solids form a three-dimensional crystal lattice in one of seven crystal systems. The smallest unit of crystal structure is called a unit cell. Amorphous solids lack a crystalline structure and their atoms are in a random pattern. Glass is an example of a form of amorphous solid called a supercooled liquid.

3. The fourth state of matter is an ionized gas composed of atoms stripped of their electrons, so plasma consists of a mixture of electrons and positive ions. This usually happens at very high temperatures. Plasmas behave like gases in some ways. Partial plasmas are created in florescent lights, neon signs, flames, and lightning. The free electrons lose energy when they collide with other particles and are recaptured by the gas ion. Plasmas formed in this way are very weakly ionized.

   Highly ionized plasmas form within stars or as a result of nuclear fusion. Much of the research for fusion reactors involves containing high temperature plasmas by magnetic fields.

UNDER PRESSURE, page 30

Gas pressure is the result of billions of gas particles colliding with an object at the same time. Atmospheric pressure occurs as the air exerts pressure on objects because gravity holds these gas particles close to the earth. Atmospheric pressure is greater closer to the surface because air is more dense at lower elevations.

Weather is greatly influenced by differences in air pressure as a result of the movement of great air masses. Low pressure areas result in stormy weather. Fair weather is associated with high pressure areas.

Air pressure is measured with instruments called barometers. The SI unit of pressure is the pascal. Air pressure at sea level is normally around 101.3 kilopascals. Other units of pressure are millimeters of mercury, and atmospheres.

One millimeter of mercury (1mm Hg) is the amount of air pressure needed to support a column of mercury 1 mm high and is a unit developed when the first mercury barometers were developed.

One standard atmosphere (1atm) is the pressure needed to support 760mm of mercury in a mercury barometer at 25 degrees C, the average atmospheric pressure at sea level.

An aneroid barometer is a sealed, flexible container with a partial vacuum inside. A needle and spring system connected to the container reveal differences in pressure. When the pressure changes, the container expands or contracts accordingly and the needle moves across a scale.

Pressure measurement are important in chemistry when determining volumes of gases at different pressures and temperatures.

MERCURY BAROMETER

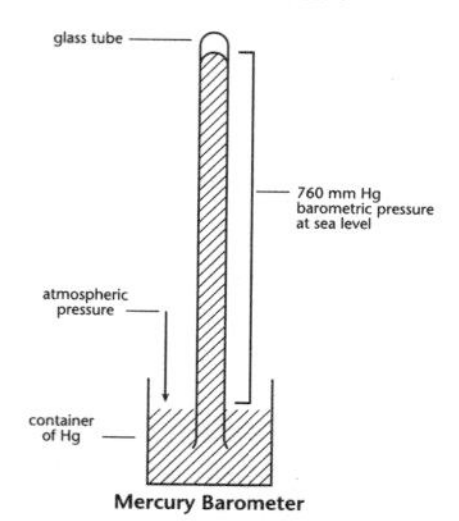

SOLIDS AND THE SEVEN CRYSTAL SYSTEMS, page 31

Drawings will vary.

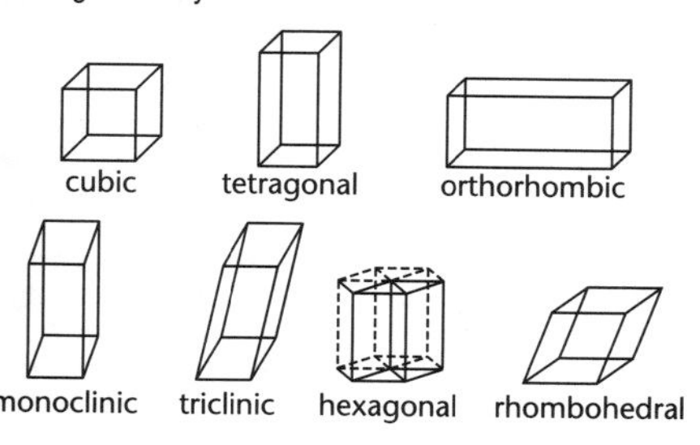

CRYSTAL MODELS, page 32

1. Monoclinic, 2. Triclinic, 3. Hexagonal, 4. Cubic

AIR FORCE DEMONSTRATION, page 33

1. The flexible plastic bends inward as the bottle partially collapses.
2. The gases inside the uncapped bottle heated and expanded when the bottle was placed in the boiling water. Because the bottle was uncapped, some of the expanding gases exited the bottle through the open top. When the bottle was capped, its sides caved in because the air pressure was now greater outside the bottle than inside.
3. To allow the expanding gases to escape as their particles gained kinetic energy.
4. The bottle would have swelled up from the increased gas pressure.
5. The sound of air rushing in is often heard: the same sound made when opening a vacuum sealed container. The plastic bottle usually resumes its normal shape, because returning air equalizes the air pressure once again.
6. When the temperature inside the bottle increased, the gas particles gained kinetic energy and the volume of the gases increased. Some of the more active gas particles escaped the bottle. When the bottle was capped and the gases inside began to cool, a pressure differential was created between the inside and outside of the bottle. Atmospheric pressure was greater outside, causing the bottle to partially collapse.

HOW DO HOT AIR BALLOONS WORK?, page 34

Every balloon rises because the gases inside the balloon are much lighter that the surrounding air, whether the gas inside the balloon is helium, hydrogen, or hot air. Hot air rises because it is only about half the density of cold air. Usually the higher the balloon goes, the colder the surrounding air and the more effectively the hot air inside the balloon becomes.

The canopy of a hot air balloon is inflated on the ground using a large gas burner and a fan. The fan blows the hot air into the canopy and the structure slowly rises to a vertical position, held down by an attached basket called a gondola and ropes attached to the canopy. The burners are attached to the gondola. An opening in the bottom of the canopy receives the heated air as it rises from the burners. Propane tanks in the gondola usually supply the burners with fuel.

When the air in the canopy is heated enough to create the required amount of lift, the ground crew releases the balloon and it rises. The balloon moves horizontally via wind currents and maintains its altitude through frequent use of the burners, which the pilot turns on and off to gain the desired altitude. Sand bags can be used as ballast and may be emptied or dropped to gain altitude quickly.

The balloon descends as the air cools inside the canopy or when the pilot vents hot air through a rope-valve arrangement at the top of the balloon canopy.

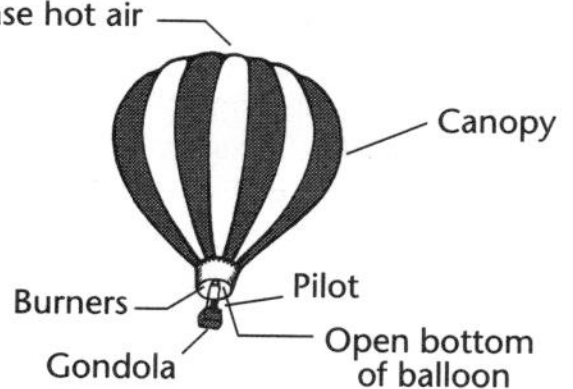

THE QUANTUM MECHANICAL MODEL OF THE ATOM, page 34

The quantum model does not define the exact path of an electron like the solar system model explains a planet's orbit. The quantum model shows the probability of finding an electron within a shaped electron cloud. Each electron cloud shape can be calculated from an equation called an atomic orbital. The arrangement of electrons around the nuclei of atoms is called the electron configuration.

There are three rules for this configuration. 1) Aufbau principle: electrons enter orbitals of lowest energy first. 2) The Pauli exclusion principle: an atomic orbital can describe no more than two electrons, which must have opposite spins to occupy this orbital together. Spinning of electrons is a quantum property and is either clockwise or counter-clockwise. 3) Hund's rule: when electrons occupy orbitals of equal energy, one electron enters each orbital until all the orbitals contain one electron with spins parallel (all spinning the same direction and none are paired up). Second electrons add to each orbital, pairing up so their spins are in the opposite direction.

De Broglie predicted that all matter exhibits wavelike motion, though this is not easy to detect. Because mechanics is the study of the motion of bodies, the quantum model became known as the quantum mechanical model, which can now be described mathematically.

Heisenberg stated it is impossible to know exactly both the position and velocity of a particle at the same time, which further confirmed the wavelike properties of matter in motion.

When an element or molecule is heated, the atoms absorb energy, then lose it by emitting light. Passing the light given off through a prism gives the atomic emission spectrum of that element, which is unique. This emission spectrum of an element consists of a series of bands of light that are characteristic of that element. It is possible to identify any element by comparing the emission spectrum of a substance with known emission spectrums of elements.

EVEN GASES MUST OBEY THE LAW, page 35

1. The pressure and volume of a fixed mass of gas are inversely related. If one decreases, the other increases.
2. The volume of a gas at a constant pressure is directly related to it's Kelvin temperature.
3. A combination of Boyle's, Charles', and Gay-Lussac's laws.
4. The total pressure in a mixture of gases is equal to the sum of the partial pressures of each gas present.
5. Gas particles moving from an area of greater concentration to an area of lesser concentration.
6. When gas escapes through a tiny hole in a container of gas.
7. The pressure of a fixed volume of gas is directly related to its Kelvin temperature.
8. The smaller the formula mass of a gas, the greater its rate of diffusion.
9. (R) = 0.0821 (L x atm)/(K x mol).
10. Relates the moles of a gas to its pressure, temperature, and volume $PV = nRT$.
11. The contribution each gas in a mixture makes to the total pressure.

ATOMS AND ELECTRONS, page 36

1. The height of the wave from the origin to the crest.
2. A region in space where there is a high probability of finding an electron.
3. All forms of radiation, such as radio waves, microwaves, infrared or heat waves, x-rays, light.
4. The energy level of an electron is the region around the nucleus where it is likely to be moving.
5. The number of wave cycles to pass a given point per time unit.
6. An electron in its lowest energy level.
7. Wavelength frequency in units per second.
8. Electrons called photoelectrons are ejected by metals when light shines on them.
9. Light quanta, quanta of energy that behave as particles.
10. The amount of radiant energy absorbed or emitted by a body is proportional to the frequency of the radiation. Planck's constant is a number used to calculate that value.
11. The amount of energy required to move an electron from its present energy level to the next higher one.
12. The colors produced when light is passed through a prism: a range of wavelengths of light.
13. The distance between the crests of a wave.

DON'T JUST WAVE, DRAW WAVES, page 37

1.

2.

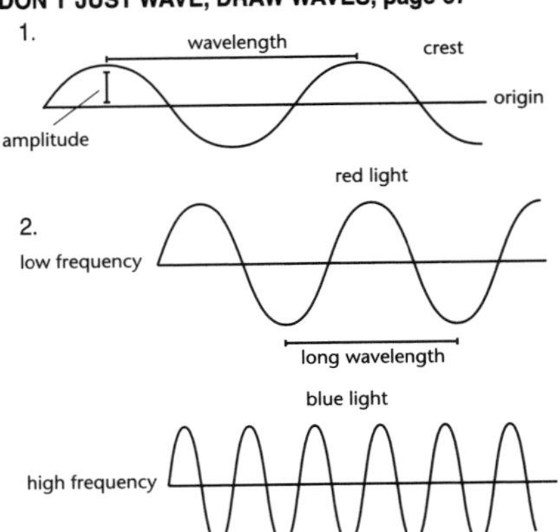

3. **Key:** R = Red; O = Orange; Y = Yellow = G = Green; B = Blue; I = Indigo; V = Violet

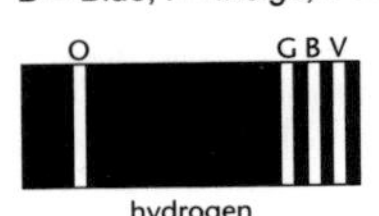

hydrogen

sodium

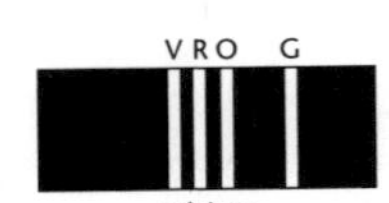

calcium

4.

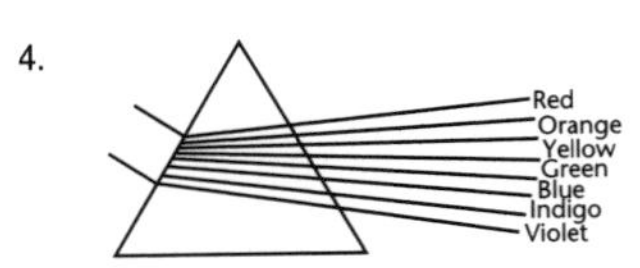

CONSTRUCTING THE PERIODIC TABLE, page 38

Dmitri Mendeleev was a master of inductive reasoning, formulating the periodic law from the known details concerning the elements. He took the information concerning the atomic and molecular masses of the elements and noticed the relationship between certain groups of them.

He listed the known elements in several vertical columns based upon their atomic mass and noticed a regular (periodic) repetition of their chemical and physical properties. He constructed a rudimentary table so that the elements with the most similar properties were side by side: the first periodic table. Blank spaces were left in the table where there were no known elements with the appropriate properties. Where an atomic mass range was missing (suggesting an undiscovered element) he was able to predict the properties of that yet unknown substance.

When the missing elements were discovered, they usually had those properties Mendeleev predicted. One element, when discovered, was calculated to have a density other than Mendeleev had predicted. He was so confident that he asked the discoverer to recalculate the density and when it was done, Mendeleev was found to be correct.

Henry Moseley, a young British physicist working with Ernest Rutherford, determined the atomic change of the elements and this became known as the atomic number. Using Mendeleev's strategy, he arranged the elements in a table, but using his newly discovered atomic numbers instead of the atomic masses of the elements. This made predictions concerning the properties of the elements even more precise. Today his system is the one used in the periodic table. Moseley also made predictions concerning the properties of undiscovered elements and these proved correct when the elements were later discovered. Unlike Mendeleev, who lived for 73 years, Moseley was unable to continue working on his early ideas. He was killed at the age of 28 in World War I.

DESCRIBING THE MODERN PERIODIC TABLE OF THE ELEMENTS, page 38

The elements are arranged in seven horizontal rows in order of increasing atomic number, from left to right. These horizontal rows are called Periods. The vertical columns are called Groups, and there are eighteen of these.

There are two additional rows of elements that are usually placed beneath the basic table. One of these rows is from atomic number 57 (Lanthanum) to 70 (Ytterbium) and is called the Lanthanide Series. The other row is from atomic number 89 (Actinium) to 102 (Nobelium) and is called the Actinide Series.

Each element throughout the table is identified by its chemical symbol placed in a block, with the atomic number at the top center or left of the block, and the average atomic mass in the upper left or lower center of the block. Other information in this block includes the element's name, state, and sometimes density, state, boiling and melting points, etc.

Each Group is identified by a number and the letters A or B. Groups IA–7A and Group O make up the representative elements, and each group is different in their general chemical and physical properties, though elements within respective groups are similar in their properties.

The nobel gases, Group O, have their outermost s and p sublevels filled (electrons) and are sometimes called the inert gases because they do not react easily.

The Group A elements (representative elements) have their outermost s and p sublevels partially filled. The group number equals the number of electrons in the outermost occupied energy level.

The transition elements are the Group B elements and have their outermost s sublevel and nearby d sublevels containing electrons. The inner transition elements have electrons in their outermost s sublevel and the nearby f sublevel. They are characterized by the filling of f orbitals.

The periodic table is divided into blocks that correspond to the sublevels filled with electrons. There is the s block, the p block, the d block, and the f block.

GOING DEEPER INTO THE PERIODIC TABLE, page 39

1. An atom does not have a clearly definable boundary to judge its size, but one way of estimating the relative sizes of atoms is the covalent atomic radius, half the distance between the nuclei of two atoms in a diatomic molecule. Moving down a group in the periodic table, atomic size generally increases, and size generally decreases from left to right across a period.
2. The energy required to remove an electron from a gaseous atom is the ionization energy. In general moving down a group of the periodic chart, the first ionization energy decreases (the outermost electron is further from the nucleus because the size is larger). For the representative elements, the first ionization energy generally increases from left to right across a period.
3. This is the energy charge that accompanies the addition of an electron to a gaseous atom. This affinity generally increases from left to right across a period and decreases moving down a group. Electron affinities generally decrease with increasing atomic size.
4. Positive ions (cations) are always smaller than the neutral atoms that they are formed from, and negative ions (anions) are always larger, because the nuclear attraction between atoms is less for an increased number of electrons. Going from left to right across a row, when the ions are arranged on the periodic table, there is a gradual decrease in the size of positive ions. Once you get to the negative ions, they continue to get smaller in size from left to right. The atomic radius with both cations and anions increases moving down each group.
5. This is the tendency of an atom to attract electrons to itself when chemically combined with another element. Going from left to right across a period, the electronegativity increases for the representative elements. Ordinarily, the electronegativity decreases as we move down a given group. The nonmetallic elements, except for the noble gases, have high electronegatives and the transition metals are variable. Charts can be made summarizing these trends on the periodic table.

LOOKING AT THE GROUPS, page 40

1. These are metals with low densities and melting points, like sodium and potassium. They react violently with water, producing metal alkalis and are not found pure in nature. When pure, these soft metals must be kept from oxygen and water.
2. These are harder and less reactive than the alkali metals, but are still not found in pure form. They tarnish quickly in air, producing a protective oxide coating that allows magnesium and beryllium to be used as low density structural materials.
3. This group contains metals like aluminum, and no-metals like boron. Lightweight, strong aluminum forms a tough outer oxide and will not react with water. Aluminum is fairly common, but not in its pure form. Some of the elements in this group are quite rare.
4. This group contains metals, metalloids, and no-metals. Carbon is found as diamonds, coal, and graphite, and is essential for life. Organic compounds are named for the presence of carbon. Silicon is the second most abundant element on earth, and is a semiconductor important to electronics. Lead and tin are also important to industry.
5. This group ranges from abundant gases to metals, with nitrogen and phosphorus being essential to life, as vital components of DNA and ATP. The atmosphere is 80% nitrogen. Arsenic, antimony, and bismuth are important metals to industry.
6. Oxygen and sulfur are very abundant and are both vital to living organisms as well as industry. Selenium is important in making photoelectric cells and light sensitive materials, like those used as light sensors and photocopying. Oxygen is the most abundant element on earth in its molecular and oxide forms.
7. This group is all non-metal. Free halogens are very reactive. Most form important salts, like sodium chloride. Many salts of fluorine, chlorine, and iodine are important for our health. The chlorides and bromides are important in industry.

PERIODIC PROPERTIES, page 41

Part A: **1.** I, **2.** C, **3.** I, **4.** I, **5.** I, **6.** D
Part B: **7.** D, **8.** D, **9.** D, **10.** I, **11.** I, **12.** I. **13.** I

ELECTRON CONFIGURATIONS, page 42

Element	Orbital Filling 1s	2s	2px	2py	2pz	3s	Electron configuration
H	↑						$1s^1$
He	↓↑						$1s^2$
Li	↓↑	↑					$1s^2\ 2s^1$
Be	↓↑	↑	↑				$1s^2\ 2s^2$
B	↓↑	↓↑	↑				$1s^2\ 2s^2\ 2p^1$
C	↓↑	↓↑	↑	↑			$1s^2\ 2s^2\ 2p^2$
N	↓↑	↓↑	↑	↑	↑		$1s^2\ 2s^2\ 2p^3$
O	↓↑	↓↑	↓↑	↑	↑		$1s^2\ 2s^2\ 2p^4$
F	↓↑	↓↑	↓↑	↓↑	↑		$1s^2\ 2s^2\ 2p^5$
Ne	↓↑	↓↑	↓↑	↓↑	↓↑		$1s^2\ 2s^2\ 2p^6$
Na	↓↑	↓↑	↓↑	↓↑	↓↑	↑	$1s^2\ 2s^2\ 2p^6\ 3s^1$
Mg	↓↑	↓↑	↓↑	↓↑	↓↑	↓↑	$1s^2\ 2s^2\ 2p^6\ 3s^2$

TERMS OF IONIC BONDING, page 43

1. N, **2.** E, **3.** K, **4.** F, **5.** H, **6.** C, **7.** J, **8.** D, **9.** I, **10.** A, **11.** B, **12.** M., **13.** G, **14.** L

INVESTIGATING IONIC BONDS, page 44

1. Chemical bonds result from the transfer of valence electrons between pairs of electrons or their sharing. This brings the atoms together as a compound.
2. Valence electrons are the electrons in the highest occupied energy level of an element's electrons, These are usually the only electrons used in the formation of chemical bonds. Knowing their number and behavior helps predict and explain chemical reactions.
3. Their valence electrons are in a stable energy configuration and, except for helium, have eight electrons in their highest energy level. They tend to hang on to their electrons, not sharing or giving any up, and not needing any more to stabilize.
4. Electrostatic forces, the forces of attraction that bind ions together with opposite changes, create ionic bonds between anions and cations. For these to form, the number of positive and negative ions must be equal in number. Compounds formed in this way are called salts.
5. Salts are electrically neutral and like sodium chloride, are usually crystalline structures at room temperature. The particles that make up these crystals are arranged in repeating three-dimensional patterns. The ions are very strongly attracted to each other in this pattern and result in high melting points for the compounds. When in a molten state, ionic compounds conduct electricity. Those that may be dissolved in water also conduct an electrical current. When dissolved, the ions are free to move about.
6. It is believed that a metal consists of closely packed cations surrounded by mobile valence electrons that drift from one part of the metal to another. The metallic bond consists of the attraction of the free valence electrons with the positive metal ions. This theory of a metallic bond explains the electrical conductivity of metals. The ductility and malleability of metals is due to the insulating ability of valence electrons between the metal cations. The structure of metal crystals is very simple.

PUT THE METAL TO THE METAL, page 45

Answers will vary. Some common alloys are:

1. Brass: 60% copper, 39 percent zinc, 1% tin. Used in gears, plumbing fixtures, lamps and household items, etc.
2. Cast iron: 96% iron, 4% carbon. Used in castings for items such as wood stoves, heavy pots and skillets, etc.
3. Coin silver: 90% silver, 10% copper. Used in medals, awards, and coins.
4. Duraluminum: 94.5% aluminum, 5% copper, 0.5% magnesium. Used in making aircraft skin and parts, lightweight bicycles and machines, boats and ship construction.
5. Gold (18 Carat): 75% gold, 25% copper and silver. Used in gold rings, chains, and other jewelry.
6. Pewter: 85% tin, 7.3% copper, 6% bismuth, 1.7% antimony. Used in dishes, tableware, decorator items.
7. Solder (plumber's old formula): 67% lead, 33% tin. Used for soldering joints in plumbing, electrical, and metal work.
8. Stainless steel: 80.6% iron, 18% chromium, 0.4% carbon, 1% nickel. Used for cutlery, tableware, and large appliances.
9. Steel: 99% iron, 1% carbon. Used as structural material for many items.
10. Sterling silver: 92.5% silver, 7.5% copper. Used for tableware, jewelry, decorator items, and picture frames.
11. Spring steel: 98.6% iron, 1% chromium, 0.4% copper. Used for springs and cutting tools.
12. Surgical steel: 67% iron, 18% chromium, 12% nickel, 3% molybdenum. Used for medical/dental implants and wires and in making surgical tools and clamps.

SETTING EXAMPLES, page 46

Answers will vary. Some possible answers are listed.

1. a) Bromine: A dense reddish-brown liquid with a sharp odor, used in making photographic emulsions.
   b) Chlorine: A greenish-yellow toxic gas, used in bleach and as a water purifier for pools.
   c) Fluorine: A greenish-yellow toxic gas, used in compounds to purify water and added to toothpaste to protect teeth.
   d) Hydrogen: Colorless, odorless gas, combined to form water, formally used in balloons, very common ingredient of many acids.
   e) Iodine: Gray-black solid, dissolved in alcohol and used as a antiseptic.
2. a) Ammonia: Colorless gas, strong odor, mixed with water and used as a cleanser.
   b) Carbon dioxide: Colorless gas, essential for plant growth, exhaled by animals, absorbs heat, used in fire extinguishers, dry ice.
   c) Carbon monoxide: Colorless, highly toxic gas produced by burning, major component of cigarette smoke and auto exhaust
   d) Hydrogen peroxide: Colorless, unstable liquid, used as rocket fuel, bleach, antiseptic.
   e) Nitrous oxide: Colorless, sweet-smelling gas used as an anesthetic.
3. a) Sodium chloride: Table salt, used to flavor food, food preservative, etc.
   b) Copper: Shiny reddish colored metal, oxidizes to a dull brownish red, used in electrical devices, cookware, wiring, coins, sculpture, construction.
   c) Calcium carbonate: A hard, gray to white substance (limestone). One of the more common minerals on earth, used in construction, works of art, landscaping.
   d) Magnesium sulfate: Clear crystals, epsom salt, used as a cathartic in medicine, industry.
   e) Barium sulfate: A white pigment for manufacturing, used in medicine for x-ray tracking inside gastrointestinal tract.

BONDS: COVALENT BONDS, page 47

1. A single, covalent bond is formed when a pair of electrons is shared between two atoms. An example is with two hydrogen atoms. Each hydrogen atom has one electron and they form a single covalent bond, sharing the two electrons and getting the stable electron configuration of helium.
2. As suggested by the names, double covalent bonds involve two shared pairs of electrons and triple covalent bonds involved three shared pairs. The carbon dioxide molecule contains two oxygen atoms that each share two electrons with carbon to form two carbon-oxygen double bonds. Nitrogen forms a triple covalent bond. Each nitrogen has five valence electrons and each needs three more electrons. A molecule of nitrogen ends up with three shared pairs and each nitrogen has one unshared pair of electrons.
3. This type of bonding occurs when one atom contributes both bonding electrons in a covalent bond. An example is with carbon monoxide. Carbon is four electrons short and oxygen is two electrons short. Two covalent double bonds form and one coordinate covalent bond forms when oxygen donates one of its unshared pair of electrons to carbon. The coordinate covalent bond acts like any other covalent bond once it is made.
4. This occurs when two different atoms are joined by a covalent bond and the bonding electrons are unequally shared. The atom with the stronger electron attraction acquires a slightly negative charge and the atom that is less electronegative acquires a slightly positive charge.
5. When the atoms in the molecule are the same, the bonding electrons are shared equally, so there is no polar effect to the covalent bonds.

THE VOCABULARY OF COVALENT BONDS, page 48

1. A molecular orbital whose energy is higher than that of the atomic orbitals from which it is formed.
2. The energy required to break a single bond.
3. A molecular orbital whose energy is lower than that of the atomic orbitals from which it is formed.
4. A molecule that has two poles.
5. Occurs when polar molecules are attracted to each other.
6. The weakest of all molecular interactions, thought to be caused by the motion of electrons.
7. Several atomic orbitals mix to form the same number of equivalent hybrid orbitals.
8. Attractive forces in which a hydrogen that is covalently bonded to a very electronegative atom is also weakly bonded to an unshared electron pair of an electronegative atom in the same molecule or a nearby molecule.
9. When two atoms combine, their atomic orbitals overlap to form molecular orbitals.
10. Very stable substances in which all the atoms are covalently bonded to each other.
11. Substances that show a relatively strong attraction to an external magnetic field.
12. The bonding electrons are most likely to be found in sausage-shaped regions above and below the nuclei of the bonding atoms.
13. One end of the molecule is slightly negative and the other end is slightly positive.
14. When two or more equally valid electron dot structures can be written for a molecule.
15. Formed when two atomic orbitals combine to form a molecular orbital that is symmetrical along the axis connecting two atomic nuclei.
16. Chemical formulas that show the arrangement of atoms in molecules and polyatomic ions.
17. In methane molecules, the hydrogen/carbon angles are all 109.5 degrees, three dimensions of a geometric solid.

18. The pairs of valence electrons that are not shared between atoms, also called lone pairs and nonbonding pairs.
19. The weakest attractions between molecules.
20. Because electron pairs repel, molecules adjust their shapes so that the valence electron pairs are as far apart as possible.

COVALENT VERSUS IONIC COMPOUNDS: A COMPARISON CHART, page 49

Covalent compounds
Type of Bond - Sharing of electrons between atoms
Elements - Non-metallic
Physical State - Solid, liquid, or gas
Melting point - Low
Solubility in water - Variable, high to low
Conductivity in aqueous solution - Poor to non-conducting
Ionic compounds
Type of Bond - Transfer of electrons between atoms
Elements - Metallic and non-metallic
Physical State - Solid
Melting point - High
Solubility in water - Usually high
Conductivity in aqueous solution - Good conductor

VOCABULARY: WATER AND SOLUTIONS, page 50

1. N, **2.** P, **3.** A, **4.** L, **5.** J, **6.** K, **7.** R, **8.** E, **9.** H, **10.** T, **11.** D, **12.** F, **13.** I, **14.** S, **15.** M, **16.** O, **17.** C, **18.** Q, **19.** B, **20.** G

MOLECULES: STAYING IN SHAPE, page 51

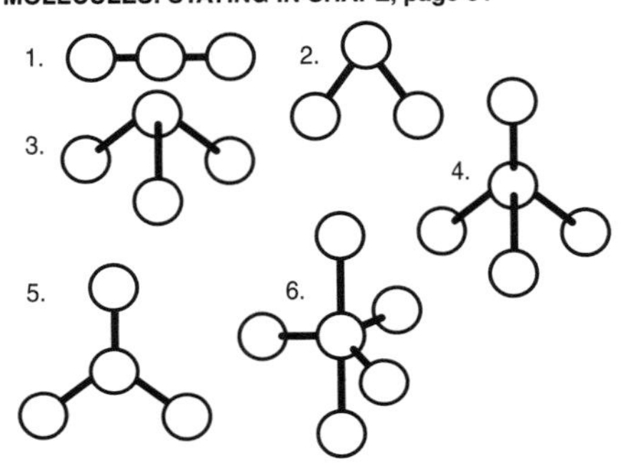

SOLUBILITY OF SOLIDS AT DIFFERENT TEMPERATURES, page 52

1. Wording of the answers will vary, but should reflect the graph results.

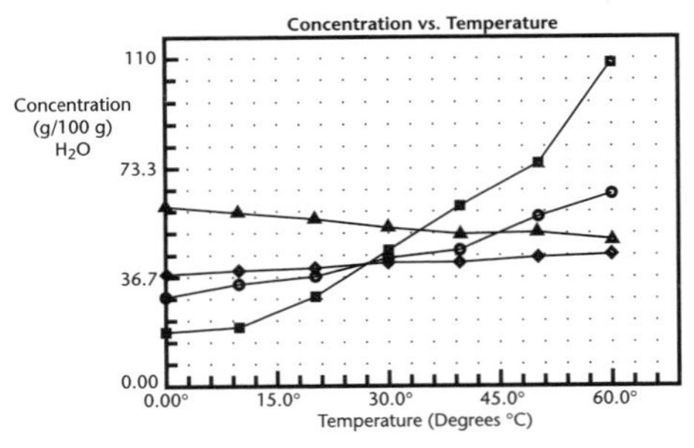

2. The graph should indicate greater solubility of gases as the temperature decreases.

SOLUTIONS TO SOLUTIONS, page 53

1. Changes in solubility are affected by differences in temperature and pressure. The amount of gas dissolving in a liquid is proportional to the pressure of the gas and the solubility of gas in liquid decreases with increasing temperature. Nonpolar liquids tend to be insoluble in water. In general, the higher the temperature of water, the greater the solubility of solids. The rate something dissolves also depends on the size of the dissolving particles.
2. This process is used to separate the components of a solution in order to identify them. In paper chromatography, colored substances can be separated in the following manner: A drop of the of filter paper. The filter paper is placed so the bottom is barely touching a chosen solvent. Capillary action causes the solvent to move up the paper and through the drop of colored substance. If the components of the substance in the drop have different affinities for the paper, they will migrate up the traveling solvent at different rates and separate out. Gas and solid solvents are used in other forms of chromatography, especially in chemical laboratories and research facilities.
3. Molarity is the number of moles of a solute dissolved in one liter of solution. The volume is the total volume of the solution, not just the solvent. Molarity is equal to the number of moles of solute divided by the number of liters of solution.
4. You never add water to acid because the heat generated by the smaller amount of water encountering the greater amount of acid when you begin adding causes the solution to sputter and splash. Adding acid to water, while stirring, insures that the sudden heat build-up doesn't occur.
5. Percent is parts per hundred. When both solvent and solute are liquids, you can make the solution by measuring volumes. The concentration of the solute is expressed as a percentage of the solution by volume. If 20 ml of solute A is added to solvent B to make a total volume of 100 ml, the final solution is 20% A by volume.
6. These are properties depending on the number of particles dissolved in a given mass of solvent. Examples are vapor pressure lowering, boiling point elevation, and freezing point depression.
   A common example of freezing point depression occurs when we add salt to ice on streets. The ice melts because the freezing point of the salt water solution formed is lower than that of pure water.

SEARCHING FOR SOLUTIONS, page 54

1. The difference in temperature between the boiling points of a solution and of the pure solvent.
2. A solution that contains a large amount of solute.
3. A measure of the amount of solute dissolved in a given quantity of solvent.
4. A solution that contains only a small amount of solute.
5. The difference in temperature between the freezing points of a solution and the pure solvent.
6. At a given temperature the solubility of a gas is directly proportional to the pressure of the gas above the liquid.
7. Liquids that are insoluble in each other.
8. Liquids that will dissolve in one another.
9. The number of moles of a solute dissolved in one kilogram of solvent.
10. The ratio of the moles of solute in solutions to the total number of moles of both solvent and solute.
11. Contains the maximum amount of solute for a given amount of solvent at a constant temperature.
12. The amount of substance that dissolves in a given amount of a solvent at a given temperature to produce a saturated solution.
13. A solution that contains more solute than it can theoretically hold at a given temperature.
14. A solution that contains less solute than a saturated solution.

MATCHING REACTION RATES, page 55

1. B, **2.** C, **3.** D, **4.** G, **5.** K, **6.** A, **7.** I, **8.** J, **9.** H, **10.** E, **11.** F, **12.** L

CHEMICAL REACTION RATES, page 56

1. The arrangement of atoms at the peak of the activation energy barrier.
2. The minimum energy colliding particles must have to react.
3. The maximum amount of energy that can be coupled to another process to do useful work.
4. When forward and reverse reactions are taking place at the same time.
5. Where reactants are converted to products in a single step.
6. The ratio of product concentrations to reactant concentrations, with each concentration raised to a power given by the number of moles of that substance in the balanced chemical equation.
7. Reached in a reaction when the relative concentrations of reactants and products are at equilibrium.
8. If stress is applied to a system in a dynamic equilibrium, the system changes to relieve the stress.
9. Reactions that do not give products under the specified conditions.
10. An expression relating the rate of a reaction to the concentration of reactants.
11. Includes all the elementary reactions of a complex reaction.
12. When the conversion of reactants into products and the conversion of products into reactants occur simultaneously.
13. A proportionality constant relating the concentrations of reactants to the rate of the reaction.
14. The stable state of a substance at 25 degrees Celsius and one atmosphere.

EXPLAINING CHEMICAL REACTION RATES, page 57

1. The activation energy barrier is the energy value that must be exceeded for a reaction to take place. The colliding particles must have enough energy to make or break bonds and the amount of energy to make this happen has to be above a certain amount. The amount that must be exceeded is the activation energy barrier.
2. For any reaction involving two or more reactants, enough particles must collide often enough with enough energy or a reaction will not take place.
3. The law of disorder states that particles move spontaneously in the direction of maximum chaos or disorder and the measurement of the amount of disorder in a system is called entropy. Entropy increases when a substance changes to a more chaotic state, as from a liquid to a gas, when a substance is divided into parts, and when temperatures increase. Reactions can move a substance to a higher or lower entropy state.
4. The rates of chemical reactions are usually increased by increases in temperature, a higher concentration of reacting particles, and the smaller the size of the particles involved. Smaller particles have a higher surface area. A catalyst can also increase the rate of a chemical reaction.
5. A catalyst is a substance that increases the rate of a chemical reaction without itself being consumed. A catalyst provides reactants with a reaction path of lower activation energy, allowing a larger fraction of reactants at a given temperature to form products. In catalytic converters on automobiles, platinum serves as a catalyst, allowing hydrogen and oxygen formed by the internal combustion engines to combine and form water as a waste product rather than dangerous exhaust gases. Enzymes in living systems serve as catalysts in biochemical reactions.

6. Reversible reactions are possible when forward and reverse reactions occur simultaneously. When mixed, reactants must produce products before reversal processes can begin. Once products build up, some products revert to reactants. As the reactants are used up, the forward process slows down and less products are produced. Eventually an equilibrium is reached. The equilibrium position depends on the relative concentrations of reactants and products at equilibrium. The product is favored if the equilibrium position contains a higher concentration of product. The opposite is true if the reactant concentration is higher. An irreversible reaction cannot reach an equilibrium position.

THE ACID-BASE TEST, page 58

1. These theories were proposed by Arrhenius, Brönsted and Lowry, and Lewis. Arrhenius stated that acids are compounds containing hydrogen that ionize to yield hydrogen ions in aqueous solutions. Bases are substances that ionize to yield hydroxide ions in aqueous solutions. Not all substances that contain hydrogen are acids, however, nor do all the hydrogens in an acid ionize. Very polar bonds are required for ionization.

   Brönsted and Lowry noted that Arrhenius's theory did not explain the behavior of all the compounds that behave like bases, such as ammonia and sodium carbonate in aqueous solutions. Neither of these compounds is a hydroxide. Brönsted and Lowry define an acid as a hydrogen-ion donor and a base as a hydrogen-ion acceptor. All of the acids and bases classified by Arrhenius are also considered acids and bases by Brönsted-Lowry, plus some extra bases such as ammonia in water. Water is also considered a base in the Brönsted-Lowry theory.

   Lewis considered the donation or acceptance of a pair of electrons in defining acids and bases. According to Lewis, an acid is a substance that can accept a pair of electrons to form a covalent bond, and a base is a substance that can donate a pair of electrons to form a covalent bond. Brönsted-Lowry acids and bases are also Lewis bases. A hydrogen ion is considered a Lewis acid and a hydroxide ion a Lewis base.
2. The strength of an acid is determined by how much the acids ionize in water. Strong acids completely ionize in water and weak acids only slightly. Strong bases dissociate completely into metal ions and hydroxide ions in water. Weak based do not. The terms acid dissociation constant and base dissociation constant are ratios of the dissociated forms to the undissociated forms and reflect the fraction of an acid or base in the ionized form. The terms concentrated and dilute tell how much of an acid or base is in solution. The terms strong and weak refer to the extent of ionization.
3. The pH scale is the negative logarithm of the hydrogen-ion concentration. For pH calculations concentrations are expressed in exponential form. A substance with a hydrogen-ion concentration of 0.01M is rewritten as 1×10^{-3} M and expressed as a pH of 3.0. A pH of 7 is considered neutral. Any number lower than that is acidic. The lower the number, the stronger the acid. A number above 7 is basic. The higher the number, the stronger the base. The pH scale ranges from 0 to 14.

ACIDS AND BASES, page 59

1. The ratio of the concentration of the dissociated form of an acid with the disassociated form.
2. Where the hydrogen ions are greater than the hydroxide ions.
3. Another name for a basic solution.
4. A substance that can act as both an acid and a base.
5. The ratio of the disassociated form of a base to the undissociated form.
6. Where the hydroxide ions are greater than the hydrogen ions.
7. The particle formed when a base gains a hydrogen ion.
8. The particle that remains when an acid has donated a hydrogen ion.
9. Two substances that are related by the loss or gain of a single hydrogen ion.
10. Any acid that contains two ionizable protons.
11. A base according to the Brönsted–Lowry theory.
12. A water molecule that gains a hydrogen ion.
13. A water molecule that loses a hydrogen ion.
14. The product of the concentrations of the hydrogen ions and hydroxide in water.
15. An acid that contains one ionizable hydrogen.
16. Any aqueous solution in which the hydroxide and hydronium ions are 1.0×10^{-7} mol/L.
17. The reaction in which two water molecules react to give ions.
18. An acid that ionizes completely in water.
19. A base that dissociates completely in water.
20. Any acid that contains three ionizable protons.
21. An acid that ionizes only slightly in water.
22. A base that dissociates only slightly in water.

GOING DEEPER: ACIDS AND BASES IN YOUR LIFE, page 60

Answers will vary. An examination of labels of household chemicals and medicines would be a good follow-up.

1. Used in acid-etching in art, making explosives, fertilizers
2. Used as a metal cleaner, etching in art
3. Used to etch glass in art and making frosted bulbs
4. Milk of magnesia, used to treat acid stomach
5. Used in photography to stop the developing process, vinegar
6. Used as a food preservative, vital to life as vitamin C, found in citrus fruits
7. Lye, used as a drain cleaner, soap making, in mercerizing, to make cotton shiny
8. Ammonia in water, used in cleaning floors and other surfaces
9. Used by ants as a defensive chemical in their bites

OXIDATION-REDUCTION, page 61

1. F, **2.** H, **3.** I, **4.** K, **5.** G, **6.** D, **7.** E, **8.** B, **9.** A, **10.** J, **11.** C

COMPARING OXIDATION AND REDUCTION, page 62

1. loss of all in ionic reactions: gain all electrons in ionic reactions
2. away from atom in a covalent bond: shift toward atom in a covalent bond
3. gain of oxygen: loss of oxygen
4. loss of hydrogen by a covalent compound: gain of hydrogen by a covalent compound
5. increase: decrease

6.–8. Answers may vary: Some examples are:
   Oxidation $C_2H_4 + 3O_2 \longrightarrow 2CO_2 + 2H_2O$
   Reduction $2KClO_3 \longrightarrow 2KCl + 3O_2$
   Oxidation $2H_2 + O_2 \longrightarrow 2H_2O$
   Reduction $CuO + H_2 \longrightarrow Cu + H_2O$

9. Except in metal hydrides, the oxidation number of hydrogen in a compound is always plus one.
10. The oxidation number of an uncombined element is always zero.
11. Except in peroxides, the oxidation number of oxygen in a compound is always minus two.

CHEMISTRY IN YOUR LIFE, page 63

1. Rust is the oxidation of metals like iron and steel. Water and oxygen in the environment convert the metal to metallic ions, converting the metal to iron buildings, and structures like bridges may be weakened or destroyed by corrosion. Water, high temperatures, and exposure to salts or acids speed the oxidation of metals. Countermeasures include coating the metals with sealants like oil, paint, or plastics, or covering with another metal. Electrochemical protection can be provided by placing another metal such as zinc or magnesium in electrical contact with the iron. When the iron is attacked by water or oxygen, the iron atoms lose electrons, which are immediately replaced by electrons from the other metal and the iron becomes neutral again. Slowly the other metal is rusted, not the iron. This is why iron is frequently coated with zinc.
2. They quickly form resistant oxide coatings, which makes them resistant to further oxidation.
3. They come in constant contact with water, often salt water, and are especially vulnerable to corrosion. Aluminum hulls are frequently used, ships are frequently repainted and sealed, and electrochemical protection is almost a necessity.
4. Carbon monoxide serves as an excellent reducing agent and combines with the iron atoms in hemoglobin, the molecule responsible for carrying oxygen in the blood stream. The iron atoms are reduced, preventing them from picking up the oxygen. Animals or humans suffer from a lack of oxygen, which can be fatal in a short time.
5. Sulfur dioxide inhibits the oxidation, or drying of fruits, and fruits treated with it look fresher.

GENERATING ELECTRICITY WITH CHEMICAL ENERGY, page 64

1. The electrode at which oxidation occurs.
2. A group of cells connected together.
3. The electrode at which reduction occurs.
4. The difference between the reduction potentials of the two half-cells.
5. A voltaic cell in which the electrolyte is a paste.
6. The ability of a voltaic cell to produce an electrical current.
7. Any device that converts chemical energy into electrical energy or the reverse.
8. The conversion of chemical energy into electrical energy or the reverse.
9. A conductor in a circuit that carries electrons to or from a substance other than metal.
10. The process in which electrical energy is used to bring about a chemical change.
11. Electrochemical cells used to cause a chemical change through the application of electrical energy.
12. Voltaic cells in which a fuel substance undergoes oxidation and from which electrical energy is obtained continuously.
13. One part of a voltaic cell in which either oxidation or reduction occurs.
14. In a half-cell, a measure of the tendency of a given half-reaction to occur as a reduction.
15. A tube containing a conducting solution which connects two half-cells, but prevents the separate solutions from mixing.
16. The measured cell potential when the ion concentration in the half-cells are 1.00M and gases are at STP.
17. A value of 0.00V used with other electrodes so that the reduction potentials of those cells can be measured.
18. Electrochemical cells that are used to convert chemical energy into electrical energy via spontaneous redox reactions.

ELECTROCHEMICAL PROCESSES, page 65

1. A dry cell is a voltaic cell where a graphite rod (cathode) and a zinc case (the anode) are separated by a moist paste of manganese oxide and other chemicals and water. The paste is surrounded by a paper barrier that keeps the contents from mixing freely, so no salt bridge is needed. The manganese is reduced, not the graphite rod. This type of battery is not rechargeable, and is used in many portable electronic devices from watch batteries to CD players.
2. Batteries are groups of cells connected together. Lead storage batteries are commonly used in vehicles like automobiles. Each cell contains lead electrodes or grids, one packed with spongy lead, the other packed with lead oxide. The grids are immersed in strong sulfuric acid, separated by a perforated plate. When the battery discharges, it produces enough electrical power to start the vehicle. When the car is running, the battery is recharged. Eventually these batteries fail because lead sulfate is lost from the electrodes and falls to the bottom of the cells or the cell is shorted out.
3. These are voltaic cells where a fuel undergoes oxidation and electrical energy is continuously obtained. They don't have to be recharged and operate more quietly and cheaply than an ordinary electrical generator.

   In a hydrogen, oxygen fuel cell, there are three compartments separated by two electrodes made of porous carbon. Oxygen is the oxidizer and is fed into the anode compartment. The gases diffuse through the electrodes. In the center compartment is a hot solution of potassium hydroxide. An external circuit routes electrons from the oxidation reaction at the anode to the reduction reaction at the cathode. The overall reaction is the oxidation of hydrogen to form water.

   Fuel cells are expensive and have applications in military vehicles, submarines, and spacecraft. The Apollo spacecraft on the lunar missions had hydrogen-oxygen fuel cells. It was the explosion of a fuel cell on Apollo 13 that caused on the problems on that flight.
4. The electrolysis of brine (salt water) results in the industrial production of chlorine gas, hydrogen gas, and sodium hydroxide. Chlorine ions are oxidized at the anode to form chlorine gas. Water is reduced to produce hydrogen gas at the cathode. The reduction of hydrogen produces hydroxide ions and the electrolyte in solution becomes sodium hydroxide.

UNDERSTANDING THE VOLTAIC CELL, page 65

Voltaic cells are electrochemical cells used to convert chemical energy into electrical energy by spontaneous oxidation-reduction reactions. A simple voltaic cell is composed of two half-cells, each containing a metal strip or rod immersed in a solution of its ions. Usually, one of the half-cells contains a zinc rod in a solution of zinc sulfate and the other half-cell contains a copper rod in a solution of copper sulfate.

The half-cells may be separated by a porus boundary or a salt bridge of a conducting solution such as potassium sulfate may connect the two half-cells. The salt bridge or porus boundary allows the ions to pass from one half-cell to the other, but prevents the two solutions from mixing completely.

A wire completes the circuit, which often includes a voltmeter or a light bulb. The circuit connects the zinc rod to the copper rod. The driving force of the voltaic cell is the spontaneous redox reaction between zinc metal and copper (II) ions in solution. The copper rod is the cathode, where reduction occurs and the zinc rod is the anode, at which oxidation occurs. The flow of electrons is from the zinc, where the electrons are produced. Then the electrons flow through the circuit to the copper rod, lighting the bulb or activating the voltmeter as they pass through. At the copper rod, copper ions are reduced.

To make the circuit complete, positive and negative ions move through the aqueous solutions via the salt bridge.

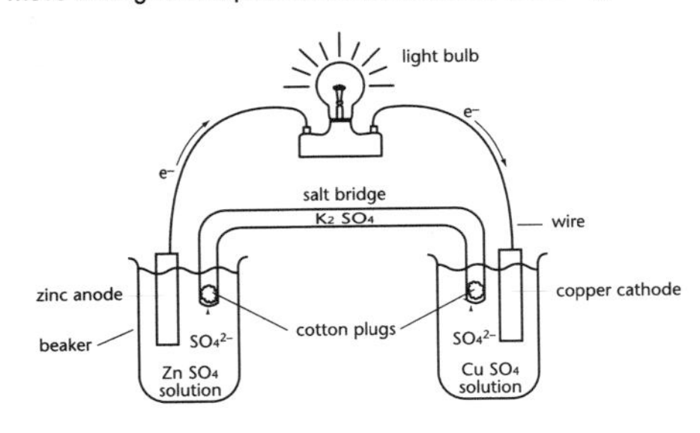

CHEMISTRY, SOURCES, AND USES OF METALS, page 66

1. These group 1A metals have very large atoms with low densities. They have low ionization energies and are very strong reducing agents. Sodium is the best known metal in this group and is a soft, shiny silvery color that is never pure in nature because it is so active chemically. It reacts violently with water and air and must be stored immersed in oils or kerosene. Sodium is a very good conductor of heat and is used in nuclear reactors to carry heat away from the reactor core. Sodium is also used in sodium lamps and in the production of chemicals. Large amounts of sodium chloride are obtained from salt beds. Sodium itself is isolated through electrolysis, which also provides chlorine gas as a by-product. The other alkali metals are only produced in small amounts.
2. These group 2A metals are similar to those in group 1A, but aren't nearly as reactive and are harder. Though less soluble than alkali salts, alkaline earth salts are very abundant as well. Magnesium and calcium are dissolved in large amounts in sea water. Calcium is the main component in the shells of sea creatures and in the skeletons of coral. Limestone is one of the most common minerals on earth, calcium carbonate, and exists as calcite, limestone, and marble. Gypsum, another important chemical used in building materials, is a calcium hydrate.

   Calcium and magnesium are important to industry. Magnesium is the only metal in this group used in the free state. Magnesium is an important alloy in making strong, low density building materials. When pure, magnesium ribbons or powder is used in flash lamps. Some alkaline earth oxides like lime and magnesium oxide are important in making mortar, firebricks, and insulation.
3. This group of metals is extremely important to industry and includes chromium, cobalt, copper, iron, gold, manganese, mercury, platinum, silver, titanium, zinc, and several others commonly manufactured. Most of these metals are ductile, malleable, and good conductors of heat and electricity. They are used for building materials, electrical devices, plating on other materials, and appliances. Iron is the second most abundant metal in the earth's crust.

   Many of these metals are alloyed together to combine properties or create new properties not present in any of the alloyed metals. Three of these metals; gold, copper, and silver, were probably the first used by humans because they can be found in the free state.

REALLY HEAVY METALS, page 67

1. Mercury and mercury salts have been used by industry for many years and these substances have ended up in the environment. Bacteria can convert mercury into highly toxic dimethyl mercury, which can enter via the skin or lungs. Mercury can cause nerve damage, brain damage, birth defects, and paralysis, as well as death. Mercury can accumulate as it moves up through the food chain, making certain fish toxic to humans. People have been known to suffer permanent damage or death from eating animals that ate seeds intended for planting rather than feed. (The seeds were treated with mercury compounds to reduce mold.)
2. Lead can cause mental retardation, kidney failure, and convulsions. Lead was a component in leaded gasoline for many years, adding lead to the environment. Lead shot from bird hunters has contaminated many lakes, ponds, and marshes. Lead from batteries has gotten into the soil.
3. Cadmium can cause kidney damage and the loss of red blood cells. It originates from improperly disposed batteries that contaminate the soil and water supplies.
4. Arsenic is required in very small amounts for reproduction. In larger amounts it causes mental disturbances, gastric distress, kidney failure, and death. Most people think of arsenic first as a poison. Arsenic is used in paints, printing, many types of poisons, and in the manufacture of semiconductors.
5. Copper is necessary as a component of several enzymes and in the formation of hemoglobin, but may cause liver damage if taken in larger amounts. Copper is widespread in electrical devices, coins, and the construction industry, and as copper sulfate in insecticides and fungicides.